autoricerca.com

AutoRicerca

No. 19, Year 2019

AutoRicerca: No. 19, Year 2019
Editor: Massimiliano Sassoli de Bianchi
Cover: Massimiliano Sassoli de Bianchi

AutoRicerca (ISSN 2673-5105) is a publication of the *LAB – Laboratorio di AutoRicerca di Base* (www.autoricerca.ch), c/o *Area 302 SA* (www.area302.ch), via Cadepiano 18, 6917 Barbengo, Switzerland.

ISBN: 978-0-244-16152-1

INDEX

autoricerca.com

WARNING

The pages of a book, whether paper or electronic, possess a peculiar property: they are able to accept whatever variety of letters, words, phrases and illustrations, without ever expressing a criticism, or disapproval. It is important to be aware of this fact when we go through a text, so that the lantern of our discernment can always accompany our reading. To explore new possibilities, we must remain open-minded, but it is equally important not to succumb to the temptation to uncritically absorb everything we read. In other words, the warning is to always subject the content of our reading to the scrutiny of our critical sense and personal experience.

The author can in no way be held responsible for the consequences of a possible paradigm shift induced by the reading of the words contained in this volume.

autoricerca.com

EDITORIAL

This nineteenth volume of *AutoRicerca* is the third of 2019. Up to now, the rhythm of publication has been of two volumes a year. With this year, we thus move the bar a bit higher, bringing the number of volumes published to three. In this way, we will have a issue 19 that will be the last published in 2019, and a future issue 20 which will be the first one of 2020. That this "numerological synchronicity" may be a good omen for the future development of *AutoRicerca*.

This volume is proposed both in Italian and English editions, to once again mark the international spirit of the journal. This time, its content is entirely dedicated to the topic of *quantum physics*, with three texts written by *Massimiliano Sassoli de Bianchi*.

The first, is a revised, updated and expanded edition of a booklet the author published in 2013, with *Adea Edizioni*, in which he offers a disenchanted view of the "mysterious" observer effect of quantum mechanics, which today we hear so much about, often in areas that have little or nothing to do with physics.

This text, upon its release, received numerous constructive feedbacks, both from Italian and Anglo-Saxon readers. So, when recently the publisher decided to simplify its catalog and no longer maintain the title, the author immediately thought not only to seize the opportunity to make a second edition, enriched in its content, but also to make this time the text freely accessible through *AutoRicerca*, which I remind you is an open access journal whose volumes in electronic format (pdf) can be downloaded free of charge from the website of the *Laboratorio di Autoricerca di Base*.

More exactly, the new text contains updated bibliographic references and the chapter on the topic of *conceptual entities* has been expanded in its final part, which now contains some necessary additional explanation, to avoid possible misunderstandings about the need to make a distinction between the "human conceptual entities" and the conjectured "conceptual entities of microphysics."

Two new chapters have also been added, one on the theme of *entanglement*, and more exactly on how to correctly understand the so-called *EPR-paradox*, and the other on the different "observer effects" that can be found outside the field of physics. Therefore, even those who have already read the previous edition of the book, by going through the book again will have the pleasure of discovering these numerous additional contents.

The second text of this volume contains the slightly revised and expanded "transliteration" of a video that the author published (first in Italian) on YouTube on April 5, 2012,[1] then on August 27, 2012 also in English,[2] entitled *Heisenberg's Uncertainty Principle and Quantum Non-Spatiality (Non-locality)*.

This writing was initially published in Italian, in 2013, by *Lulu.com*, on behalf of the author, and last year an English version of it was also uploaded as an article on the *arXiv*, the e-print service owned and operated by *Cornell University*.[3] Since its content is complementary to that of the text on the observer effect, the author thought it would be advantageous to also include it in the present issue.

Finally, this nineteenth issue of AutoRicerca contains a short article in which the possibility of *spontaneous self-teleportation* of a human body is explored,[4] as allowed by the quantum laws. The author wrote it following a question that was addressed to him by a science fiction writer, trying to offer an answer which is both

---

[1] *http://youtu.be/nN3BWe4LanQ.*

[2] *https://youtu.be/9C3vtVADL1o.*

[3] *https://arxiv.org/abs/1806.07736v1.*

[4] The article was uploaded on the *arXiv* in 2017, in its popular physics section: *https://arxiv.org/abs/1712.08465.*

qualitative and quantitative, that is, also making an explicit calculation of the probability of such an event. Being the content of the article perfectly relevant to the concepts already explored in the two previous texts, this offers a further and valuable reading complement. Naturally, the part where the probability of self-teleportation is explicitly calculated can only be understood by those who know the mathematical formalism of quantum physics, but much of the article remains nonetheless accessible to the general reader.

As always, I wish you a good study and an enjoyable reading.

The Editor

autoricerca.com

ABOUT THE AUTHOR

Massimiliano Sassoli de Bianchi received the Ph.D. degree in physics from the Federal Institute of Technology in Lausanne (EPFL) in 1995, with a study on temporal processes in quantum mechanics. His current research activities are focused on the foundations of physical theories, quantum mechanics, consciousness studies and quantum cognition. He carries forward interests in the field of inner research (self-research), promoting a multi-dimensional view of human evolution. He has written essays, popular science books, children's stories, and has published numerous research articles in international journals. He is the editor of the Italian journal *AutoRicerca* and currently the director of the *Laboratorio di Autoricerca di Base* (LAB), in Lugano, Switzerland. He is also a research fellow at the *Center Leo Apostel for Interdisciplinary Studies* (CLEA), at the *Vrije Universiteit Brussel* (VUB), in Belgium. For more information, refer to the author's personal website: *www.massimilianosassolidebianchi.ch.*

autoricerca.com

OBSERVER EFFECT

Massimiliano Sassoli de Bianchi

FOREWORD

This text offers a disenchanted view of the "mysterious" *observer effect* of quantum mechanics, which today we hear so much about, often in areas that have little or nothing to do with physics. It contains the transcript – revised and expanded – of a public conference that the author held in the city of Lugano, in March 2012.

The text is addressed both to lay readers in science – but nevertheless curious and willing to get intellectually involved – and to so-called "experts," who will find in the thesis here presented an advanced perspective concerning the delicate subject of observation in quantum physics, which I hope will be able to stimulate further studies, for example through the reading of the more recent works of the *Geneva-Brussel school* on the foundations of physical theories, from which this writing draws much of its inspiration (see the bibliography).

Among the non-expert readers, some may have heard of the observer effect from authors having a marked orientation towards Eastern philosophies, or New-Age movements. In these areas, it is often touted – unfortunately in an entirely uncritical way – the idea that the observer effect of quantum physics would constitute the "scientific proof" that the human minds are capable of acting directly on the material substances.

Apart from the fact that science doesn't deal with proofs, but rather with explanations and "errors hunts," and regardless of whether a mind-matter interaction would be possible or not, it is important to understand that the observer effect described by quantum physics has nothing to do with a *psychophysical* effect, but rather with a completely physical *process of creation*, which is inherent in some of our *non-ordinary* modalities of observing reality.

Therefore, this booklet can also be regarded as the symbol of a more mature dialogue between science and spirituality, so that both research fields can grasp the differences in their mutual cognitive paths, without unnecessary reductionisms and harmful simplifications; it is only through a correct perception of such differences, in fact, that a genuine dialogue and a possible cooperation may become possible.

ABSTRACT

If *quantum mechanics* is a complete theory, then according to its orthodox interpretation, no phenomenon can be such if not first observed, and reality cannot exist in the absence of observation.

To this paradoxical conclusion, known as the *observer effect*, or *measurement problem*, which gave birth to one of the most articulate intellectual debates in the history of science, *Albert Einstein* retorted, quite rightly, that the Moon continued to exist, undisturbed, even when nobody was watching it!

But to what extent can we say that our observations can create our own reality? And is it really true that quantum mechanics would have reached the same conclusions as some mystical-religious philosophies, which claim that the universe is a product of the consciousness?

The purpose of this booklet is to introduce the reader, specialist or non-specialist, to the reasons of the thorny question of the observer effect, in order to clarify the true nature of the process of observation in quantum mechanics.

We will do this by demystifying the whole thing on the basis of the realistic approach known as the *hidden measurement approach*, proposed in the eighties of the last century by the Belgian physicist *Diederik Aerts*. More precisely, through the analysis of a very ordinary physical system – a simple elastic band! – we will show how a correct understanding of the origin of *quantum probabilities* doesn't allow concluding about a hypothetical *psychophysical effect*, inherent in the process of observation.

This will lead us to consider the *true mystery* of quantum mechanics, which is not the understanding of the role of the observer-consciousness, but the genuinely *non-spatial* nature of microscopic entities, whose behavior is much more similar to that of the *abstract human concepts*, than that of the concrete objects of our daily reality.

1. INTRODUCTION

The purpose of this short manuscript is to explore in a non-technical, although conceptually accurate way, some aspects of the important theme of *observation* in *quantum mechanics*. In particular, we will try to shed some light on a key question, still quite controversial, which is the one of the very nature of the *observational process*. We will do this by answering the following question:

> Is the observation of a physical system always amenable to a process of *discovery* of a reality that was already existing, before the observation was carried out, or, in certain circumstances, can it be traced to an act of pure *creation* (or destruction), that is, to a process through which what is observed is literally *brought into existence* (or annihilated) by the process of observation as such? And if so, what is at the origin of such creative (destructive) process?

This is obviously a key question, both for the research in physics and for a broader understanding of the relationship between the human consciousness and the reality that is the subject of her/his experience. What exactly is our role as *observers-participators* of the reality in which we are immersed? Are we the discoverers of this reality or, unbeknownst to us, are we also its *co-creators*?

From the point of view of physics, these kinds of questions have emerged with the birth of one of the greatest scientific revolutions of our time: *quantum mechanics* (today more generally referred to as *quantum physics*). It is indeed in this context that in the early decades of last century a more thorough and refined investigation of the central role of the observing subject, in the characterization of the properties of a physical system, became necessary, in a way which was totally unexpected.

Indeed, the founding fathers of quantum physics did realize, during the construction of this baffling theory, that the reality of physical systems seemed to depend on the manner in which the investigators were operating on them, in the sense that it was not anymore possible to attribute certain properties to a physical system, independently of the acts of observation that it was conceivable to execute on it. From this apparently new situation, a question of purely metaphysical nature emerged, about the nature of the reality in which we live, and more specifically about the validity of the hypothesis of *realism*, which until then had been widely shared by most physicists and philosophers of science.

Roughly speaking, we can define the idea of realism as the hypothesis that "there is a reality out there," whose existence is entirely *independent* of the observing subjects, and that this reality, precisely because autonomously existing, would be knowable and describable in an *objective* way, for example through the construction of appropriate scientific theories (explanations). To put it in more suggestive terms, according to the view of realism, it would always be possible, at least in principle, *to speak about the reality regardless of the mind of the observing subject who studies and contemplates it.*

Before the advent of quantum physics, the idea of realism, at least in physics, imposed by itself, and this for one simple reason: the observing subject did not appear at any level in the physical theories. In other words, everything we knew about the physical systems and their evolution could be described independently from the existence of those who studied them: that the systems were observed or not, this did not alter in any way their properties and the way in which these properties evolved in time.

Figure 1. The Moon moves on its orbit around the Earth, regardless of the human activity on the surface of the planet.

The characteristics of the orbit that the Moon describes around the Earth, for example, remain such irrespective of the fact that terrestrial astronomers point their telescopes in order to observe it. And that is why *Johannes Kepler*, in his famous laws describing the motions of the planets, made no mention of a possible influence on these motions caused by the astronomers' activities: the planets go through the cold and quiet outer space totally careless of the bustling human activity on the surface of planet Earth!

But with the advent of quantum physics, all this suddenly changed. In fact, in the description of *microscopic* systems, physicists realized that it was no longer possible to describe these entities without mentioning in their theories the very process of observation, namely the effects that such a process could produce on the observed systems (in physics, one mostly uses the term *measurement*, instead of *observation*, but the meaning is, in ultimate analysis, exactly the same: to measure a physical quantity means, in fact, to *observe* its value in practical terms).

This strange "pitch invasion," which meant that scientists were seeing themselves – as in the mirror – represented in their own physical theories (not as the authors, but as an integral part thereof), has obviously undermined the very assumption of realism, on which rested the whole edifice of scientific inquiry, aimed at searching for an objective view of reality.

In fact, without the possibility of separating the scientists, in their role of investigating and observing subjects, from the object of their investigation and observation, how was it possible to continue to give a meaning to the very concept of reality? How was it possible to speak of reality if it could not be described independently of the thinking minds of those who were studying it?

As we will try to explain in this booklet, even though quantum physics has revealed to us some very strange and unexpected aspects of the profound nature of physical entities, particularly at the microscopic level, and although, undoubtedly, it has shown to us that it is not possible to generally describe a physical system regardless of the active role played by the observer in this description, not for this we must give up the idea of realism, i.e., the idea of a reality independent of the conscious mind activity of the observer.

To do this, however, it is necessary to abandon that form of naive realism, of a *classical* kind, which is based on the *prejudice* that the physical entities populating the world should necessarily always possess, in *actual* terms, all the properties that possibly characterize them, so that the result of whatever process of observation must necessarily always be, in principle, predictable and predetermined.

As we will show, this form of naive realism needs to be reformed in a more articulated and mature concept of realism, which sees in the *creation-discovery* binomial the key to a proper understanding of the role of the observer.

But to do this, and in order to make this text also accessible to the non-specialist readers, we will first explain what happened in the recent history of the young Western science, which has caused the observer to slip into the very structure of physical theories, and that in spite of all efforts, physicists haven't been able to "put it back into its place."

2. TWO REVOLUTIONS

At the beginning of last century, physics seemed to have reached a degree of completeness really enviable. The universe appeared to the scientists as a giant mechanism comfortably installed within the unchanging theatre of the *three-dimensional space*.

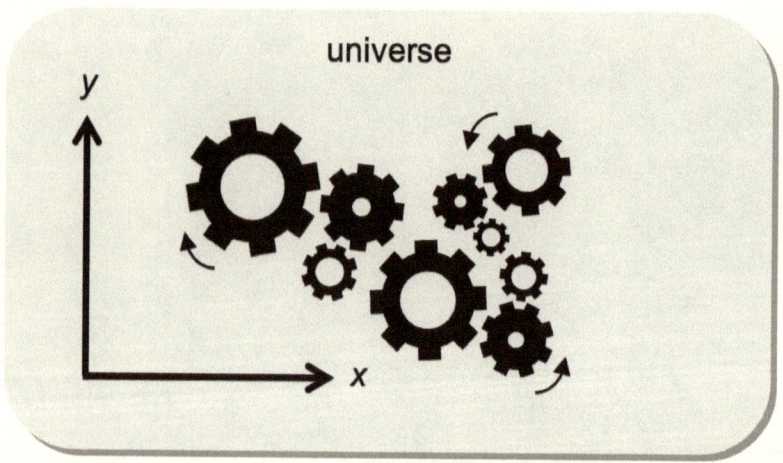

Figure 2. *A (here two-dimensional) symbolic representation of the (three-dimensional) physical space, containing the great "mechanism" of the physical reality.*

It was certainly a very complicated mechanism, about which, however, it was believed we already knew all the essential gears. The famous laws of *Isaac Newton* seemed to be able to explain all the known properties of the different material bodies, be them of a macroscopic nature, such as solids, liquids and gases, or of a microscopic nature, such as the molecules and the atoms, whose existence was at that time quite a certainty.

By analyzing for instance the collective behavior of the atomic constituents, thanks to the statistical methods devised by *Ludwig Boltzmann*, one could easily deduce the properties of the macroscopic systems, and of their various transformations, inferring them from the properties of their constituent parts, thus confirming the ancient *reductionist* assumption, according to which the whole is always equal to the sum of its parts, in the sense that the parts always make it possible to deduce and explain in a complete way the behavior and the properties of the whole.

The laws governing the electromagnetic phenomena were also known, described by the theory of *James Maxwell*, which accurately predicted not only the existence of the electromagnetic fields, emitted by the moving charges (just as gravitational fields were instead emitted by bodies with a non-zero mass), but also of the electromagnetic waves, of which light was a particular case, able to propagate as vibrations of a strange substance, immensely thin and stiff, called the *ether*, which was supposed to pervade the entire three-dimensional space.

So, in short, not without a certain conviction, it was believed that there were no more great mysteries to elucidate in relation to the physical reality, which was the foundation of any other reality, and that it was only about perfecting the different descriptions and explanations, on the basis of the laws which had been already identified.

Instead, within a few years, the whole explanatory edifice of classical physics was put in deep crisis, and had to face two great revolutions: that of *relativity*, which emerged largely from the work of *Albert Einstein* and *Henri Poincaré*, and that of *quantum mechanics*, also resulting from the work of Einstein and many other scientists, such as *Max Planck, Niels Bohr, Werner Heisenberg, Erwin Schrödinger, Wolfgang Pauli, Paul Dirac* and *John von Neumann*.

In this booklet we will not deal with the specific change promoted by relativistic theories. It should be said, however, that the discovery of relativity – first the special, then the general – did not upset the convictions of the early twentieth century physicists in the same way as the discovery of the quantum laws did. In fact, the so-called *principle of relativity*, on which rests the whole theory of Einstein, was certainly not discovered by the famous German scientist, as such a principle is as old as physics: Galileo had in fact already pointed it out (though he didn't call it that way) and described, in his admirable *Dialogue Concerning the Two Chief World Systems*, published in *1632*.

This principle states that although there are points of view about reality that obviously offer different perspectives on the phenomena, nevertheless, there is a special class of points of view (a physicist would speak here, more precisely, of *frames of reference*) which can be considered *equivalent*. These are the points of view of so-called *inertial observers*, that is, of those observers who move in the three-dimensional space at a uniform (constant) speed relative to each other.

These points of view are all equivalent in the sense that different inertial observers experiment exactly the same physical laws, and although they obviously do not measure the same physical quantities (the speeds of the bodies for example, as is known, vary depending on the relative speed of the observers; see Figure 3), there exist nevertheless simple transformations – called *Galilean transformation* – that allow to relate together the data from different inertial observers, as a universal translator would do.

What Einstein did was simply to take full advantage of the principle of relativity already identified by Galileo, discovering that Galilean transformations were only valid when the speeds involved were small compared to a *limit speed*, which was assumed to correspond to the speed of light in vacuum. In other words, the transformations of Galileo were only an approximation of more general transformations, called *Lorentz transformations*, and therefore, strictly speaking, Einstein did not invent relativity, but he reformed it.[1]

---

[1] Lévy-Leblond, J.-M. *De la matière: relativiste, quantique, interactive.* Seuil (2004).

Figure 3. *A same entity (a cat) is immobile relative to the observer on the left (i.e., with respect to the frame of reference attached to her/his body), while it is moving relative to the observer on the right. In other words, the observed speeds are not absolute, but relative to the specific point of view of each observer.*

According to these more general transformations, it appeared that there were physical quantities, which before were believed to be *invariant* with respect to the different inertial observers, which in fact were not. For as long as the speeds involved were small compared to the limit speed, the length of the objects, their inertia, the simultaneity of two events, the frequency of the ticking of a clock, etc., were all quantities that were measured (i.e., observed) by every inertial observer with apparently identical values. But as soon as the speeds involved were not anymore insignificant compared to the limit speed (the light speed in vacuum), also these quantities were able to vary according to the different points of view of the inertial observers.

So, long story short, the Einsteinian relativity had pointed out the existence of strange *generalized parallax effects*, involving physical quantities which were previously considered to be intrinsic to the objects observed. In this way, it was emphasized that the *perspective differences* between the different observers of reality were actually much more widespread than what had been thought hitherto.

Now, although the relativistic discoveries have profoundly changed the way we understand the concepts of space, time, speed, energy, etc., nevertheless, these findings did not in any way undermine the assumptions of realism. In fact, even before Einstein's relativity it was clear that any description of reality was relative to the point of view adopted: with Einstein, simply, the range of this relativization was stretched out, to include other variables.

On the other hand, nothing in the relativity theories has ever precluded that a single individual observer, from her/his particular vantage point, could describe reality in a complete way, and that her/his specific point of view could be translated into that of any other possible inertial observer of the universe. In essence, although there was a relativity of the different points of view (which was further expanded into the theory of general relativity), since these were translatable into each other through the universal translator of the *Lorentz* transformations (generalizing those of Galileo), it was always possible to affirm that there were in fact (according to relativity) a single (although manifold) description of reality, which did not depend in a strict sense on the observers contemplating it.

But the same could not be said of the quantum revolution, which changed in a much deeper and radical way our understanding of the nature of the various entities of the material universe. If relativity has certainly changed the characteristics of the spatial (and temporal) theater in which the great representation of physical reality takes place, and expanded the range of costumes available to the different actors, quantum physics has instead radically changed the very nature of these actors, and has indicated the existence of sets that could no longer be contained within the narrow confines of the ordinary three-dimensional space. And it is precisely because of this profound and troubling change, operated by quantum physics, that Feynman once said:[2]

> *There was a time when the newspapers said that only twelve men understood the theory of relativity. I do not believe that there ever was such a time. There might have been a time when only one man did, because he was the only guy who caught on, before he wrote his paper. But after people read the paper, a lot of people understood the theory of relativity in some way or other, certainly more than twelve. On the other hand, I think I can safely say that nobody understands quantum mechanics.*

We, however, will try to disprove (at least in part) Feynman's admonition, and try to really understand something about this strange and mysterious theory. But what happened exactly, at the beginning of the last century, which caused the quantum revolution to take place? It's very simple: physicists, quite unexpectedly, found themselves confronted with some experimental data that their admirable classical theories were unable to explain and predict. It is not important to enter here into the merits of these phenomena: among them, to name only the most important, we can recall the electromagnetic radiation emitted by the bodies as a function of their temperature (and more exactly, from those ideal bodies called *black bodies*), the photoelectric effect, and the colored bright lines emitted by certain ionized gases (emission spectra).

These phenomena did in fact falsify (i.e., invalidate) the existing classical theories, and it was thus necessary to fix them, or to identify new physical theories to be founded on basis yet to be identified. A number of scientists, including Planck, Einstein, Bohr, Pauli, Heisenberg, Schrödinger and Dirac, just to mention the most influential, with different timings and approaches went to work, to find a way to account for those new embarrassing experimental data, which challenged the so far known laws of physics.

Simplifying (and somewhat caricaturing) the discussion, we

---

[2] Feynman, R. P. *The character of physical law*. London: Penguin Books (1992).

can say that what they initially tried to do was to find a *mathematical model* capable of predicting the data observed in the laboratory. That is to say, in a sense, even before trying to *understand*, they tried to *predict*, i.e., to identify suitable mathematical relationships able to reproduce the values of the physical quantities which were observed (i.e., measured) in the laboratory.

Beyond all expectations, this attempt of pure mathematical modelization was a great success. Two models were initially found, very different from each other: one of an algebraic kind, based on *matrices*,[3] elaborated by Heisenberg, and another based on *differential equations*,[4] due to Schrödinger. Later on, *Dirac* and *von Neumann* showed that these two mathematical models were different only in appearance, as they were in fact both describable in the ambit of a much more general scheme, which made use of a more sophisticated kind of mathematics, based on *vector spaces*[5] of infinite dimension (the so-called *Hilbert spaces*) and *linear self-adjoint operators*[6] acting on these spaces.

In summary, in a short time it was possible to have an abstract mathematical model, very precise and efficient, able to predict almost all the new experimental data with an accuracy that, to say it all, has never been reached until then in the field of physics. But there was one problem: contrary to what always happened in the construction of physical theories, instead of identifying first what were the relevant concepts and physical quantities, defining and clarifying them on a solid operational

---

[3] A matrix is an ordered table of numeric elements.
[4] A differential equation is a mathematical relationship that links a function to its derivatives.
[5] A vector space is a mathematical structure that generalizes that of the set of three-dimensional space vectors (or of those of the two-dimensional plane), equipped with the operations of vector sum multiplication by real numbers. In the Hilbert spaces of quantum mechanics, multiplication by real numbers is replaced by multiplication by complex numbers. In other words, a Hilbert space is a vector space (possibly of infinite dimension) on the field of complex numbers.
[6] An operator is called self-adjoint when it has a specific symmetry.

base, and only then build a suitable mathematical model, this time the procedure was somehow reversed: the construction of the formal mathematical model preceded the work of clarification at the level of the physical concepts.

The consequence of all this is that physicists found themselves with a predictive tool of great power and precision in their hands, but that they couldn't fully understand, in the sense that it wasn't clear anymore what was the correspondence between the entities described in the mathematical theory and those in the physical reality. In other words, the formulation of what became known as *quantum mechanics* (now more commonly designated as *quantum physics*), posed from the beginning a serious problem regarding its interpretation, so much so that almost eighty years from the complete formulation of the theory this problem has not yet been solved, in the sense that there are still a multitude of different interpretations and formulations which, while agreeing on the experimental predictions, explain the physical content of the theory in a completely different way.

Of course, the controversial interpretative aspects of quantum mechanics are numerous, and it is certainly not possible, in the ambit of the present short essay, to list them all. Also, there is no unanimity about what would be all the conceptual difficulties that are posed by this theory. But as you can infer from the title of this booklet, we will focus here on one in particular of these difficulties – certainly not the least! – which is the one of the specific role played by the observer in the process of discovery and creation of the physical reality.

To do this, we must first understand the substantial difference between a *classical probability* and a *quantum probability*.

3. QUANTUM PROBABILITIES

As we said, one of the major conceptual challenges posed by quantum physics was to correctly interpret the physical content of the mathematical theory it referred to. Particularly sensitive was, and still is, the question of understanding the true nature of the *probabilities* involved in the theory. Indeed, contrary to classical physics, in quantum physics the concept of probability seems to play the lord and master, not only for its ubiquity, but also for the completely new meaning it takes.

Of course, also in the development of classical theories physicists had been able to become familiar with the basic concept of probability. For example, in so-called *statistical mechanics*, probabilities were used to deduce the values of the physical quantities of macroscopic bodies, such as a gas, from the properties of their atomic constituents, since the details of the individual behavior of the latter were not knowable in practice.

For example, it was not necessary to know the energy of each single molecule of an ideal gas to deduce its temperature. Indeed, it was possible to demonstrate that the gas temperature depended solely on the *average value* of the molecular kinetic energy, and to calculate an average value it wasn't necessary to possess an exact knowledge of the system, but only a *statistical* knowledge of it.

In other words, probabilities have always been the instrument used by physicists to optimally quantify their *lack of knowledge* about the specific properties of a system. The key thing to understand is that probabilities, understood here according to the classic meaning of the term, always refer to properties *already present* in the system (i.e., *already existing*). That is to say, *classical probabilities* only quantify our degree of ignorance about those *elements of reality* which in principle could be completely known.

As the understanding of this aspect is absolutely fundamental, we will make use of a very simple example, to make it as explicit and clear as possible. To this end, we consider a box containing *100 uniform elastic bands* (which in the following we will simply call "elastics") of two different *colors – black[7]* and *white* – well mixed together (see Figure 4).

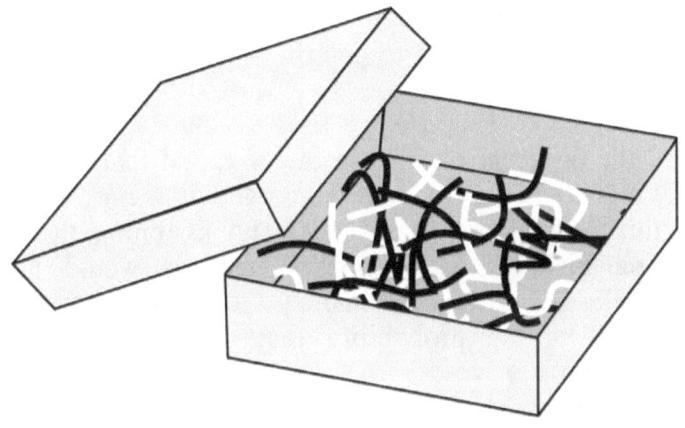

Figure 4. *A box containing 100 assorted (uniform) elastics, whites and blacks, from which, without looking, we extract a single elastic.*

Without looking, with eyes closed, we insert a hand into the box and extract one of the elastics. Since we have not looked, and we are still not looking at the elastic we have drawn, of course we are not able to determine what is its color: if black or white. We hold the elastic-entity in our hand, but we don't know its chromatic property. To put it in more scholarly terms: we are in a typical situation of *lack of knowledge* about the (chromatic) *state* of the selected elastic.

---

[7] Although the *black* in fact corresponds to an absence of color, for simplicity we will consider it, like the *white* (which is a color of high-brightness, but with no dye), as if it were a color.

To make the situation even more specific, suppose we ask the following question:

| *Is the color of the elastic drawn from the box, black?*

This question, of course, only admits two possible answers: *yes*, or *no*. It is also obvious, considering our condition of lack of knowledge, that until we don't directly observe the elastic, we will not be in a position to determine which one of these two alternatives is correct.

On the other hand, suppose we have counted, prior to the extraction, the different elastics in the box, and that we have found that it contained exactly *50 blacks* and *50 whites*. On the basis of this preliminary knowledge, and assuming that no special mechanism in the extraction process would have favored an elastic rather than another, then surely, we can calculate what is the probability that the answer to the previous question is a "yes."

In fact, as there are *50* black elastics on a totality of *100* elastics in the box, this probability is exactly *50%*. What does this mean? Simply, that if we would repeat a large number of times the extraction (by replacing each time the elastic drawn into the box), then on average, in *50%* of cases, the extraction of a black elastic would occur (and in the remaining *50%* of cases the extraction of a white elastic would occur).

Let us now come back to our single extraction. We still have the elastic in our hand (that we have not yet gazed at), and we are wondering if its color is black. All that we can say is that the *probability* that it is black, is *50%*. At this point, we can conclude the experiment and just look at, i.e., *observe*, the color of the elastic in our hand. Let us assume that we discover that it is indeed a black colored elastic.

Figure 5. *The observation of the elastic in our hand reveals that its color is black.*

This means that immediately after the observation, the probability of the elastic to be black will suddenly switch from *50%* to *100%*. What has changed, however, in this process of observation, is only the degree of knowledge of the observer, relative to the chromatic properties of the elastic. In other words, in the process of *acquisition of knowledge* by the observer, absolutely nothing special happened to the elastic: its color was black before the observer decided to look at it, and it remained black after s/he has done it.

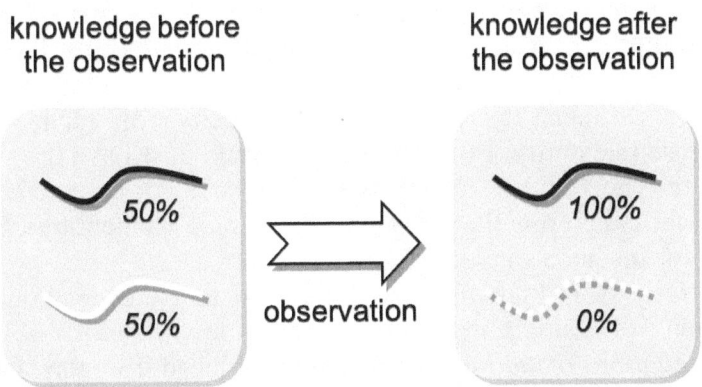

Figura 6. *In seguito all'osservazione, la conoscenza dell'osservatore circa le proprietà cromatiche dell'elastico estratto muta radicalmente e istantaneamente.*

This statement is so obvious that the mere fact of emphasizing it confers to the present discussion a sort of strange triviality. But we have to go through this triviality if we really want to understand what there is at stake when in quantum physics one speaks of the *observer effect* (or *measurement problem*), and that to that effect specific metaphysical interpretations are attached.

In summary, what we have so far highlighted is what a probability, in the classical[8] sense, corresponds to, and that a classical probability does nothing else than quantifying our degree of ignorance about a property (in our example, the color property) which is already present, that is, already *actual*, in the system considered, compatibly with the assumptions of classical realism.

So, what has changed, with respect to this explanatory framework, with the advent of quantum mechanics? What do *quantum probabilities* have that is so different compared to the classical ones? Well, first of all in quantum theory the concept of probability becomes central, in the sense that the majority of experimental results related to microscopic entities are predictable only in probabilistic terms.

This is a remarkable change, as in classical mechanics probabilities were essentially used with systems composed of a large number of components, such as for instance the molecules of a gas. In this case, it is quite natural to think of not being able to know in detail, at any instant, all the properties (such as position and velocity) of each single microscopic entity. But no physicist would ever have dreamed, before the advent of quantum physics, to describe in probabilistic terms the properties of a single elementary entity, like for instance an electron.

In fact, as is known, if at a given instant the position and velocity of a single material body are known, then by solving the *equations of motion* (which can be deduced from the famous

---

[8] In technical language, *classical probabilities* are sometimes referred to as *Kolmogorovian* probabilities, as they obey to three specific mathematical axioms that were identified by the Russian mathematician *Andrei Kolmogorov*.

Newton's laws) it is always possible to predict with certainty (i.e., with a *100%* probability) each subsequent position (and velocity) of the body. But with quantum mechanics this wasn't anymore possible. In fact, while perfectly knowing the *state* of an electron (or of any other microscopic entity) at a given instant (i.e., all the properties actually possessed by the electron at that instant), it was no longer possible to determine with certainty the trajectory that it would follow in the three dimensional space, i.e., the different positions that it would acquire in the course of time, but only calculate the probability that, if observed, the electron would be detected in a given region, in a given instant.

The initial reaction of physicists, especially Einstein, was to believe that quantum theory could not be a complete theory. In fact, if the best we were able to do was to predict the results of the different experimental observations in probabilistic terms (apart from some exceptions), this could only mean that some important information about the properties actually possessed by the microscopic entities was still missing. It was thought, in short, that there were *hidden variables*, as yet unknown, that it was necessary to identify in order to fully determine the state of the microscopic constituents, and enable in this way to predict the experimental results no longer in probabilistic terms, but with absolute certainty.

Many theoretical physicists went hunting for these fateful *hidden variables* (or properties), which would allow doing without the probabilistic description, trying to determine on which kind of *lack of knowledge* was based the probabilistic calculus of quantum mechanics. But the hunting was not successful. Not only physicists failed to discover these mysterious hidden variables, but some theorists even began to demonstrate some "inconvenient" theorems, indicating the very impossibility of such theories (which were called *hidden variables theories*), such as the famous theorems of *Gleason* and *Kochen-Specker*.[9]

---

[9] Gleason, A. M. "Measures on the Closed Subspaces of a Hilbert Space." J. Math. Mech., 6, 885–893 (1957); Kochen, S. and Specker, E. P. "The problem of hidden variables in quantum mechanics."

Faced with these difficulties, there were essentially two positions (of course, we are once again simplifying to the extreme). The majority of physicists simply lost interest in the problem. After all, the theory allowed calculating and predicting everything that was needed, although in most cases only in probabilistic terms. Thus, to discuss about the realism or anti-realism subtended by its formalism was a debate that could at best be of interest to philosophers. This extreme position, known as *instrumentalism* in philosophy of science, has been synthesized with great effectiveness by the physicist *David Mermin*, in his famous injunction:

| *Shut up and calculate!*

The remainder of physicists instead, simply decreed that the nature of quantum probabilities had to be different, in the sense that unlike classical probabilities, whose origin was clearly *epistemic*, i.e., relative to a situation of lack of knowledge, quantum probabilities were instead of an *ontic* nature, i.e., were referring to a fundamental and irreducible unknowability, genuinely present at a fundamental level in the physical reality. This position was well summarized by *Aage Bohr*, son of the famous *Niels*, who more recently has proposed to erect such an assumption to a real principle of physics, called the principle of *genuine fortuitousnes.*[10]

In other words, according to this view, quantum probabilities would not express, like classical ones, a lack of knowledge by the observer about the properties already possessed by the different physical systems, but instead a sort of mysterious *tendency* of such systems in *manifesting* their properties, *in a way which is a priori totally indeterminable*, in the course of a specific process of observation.

But even though the two extreme interpretational positions of

---

Journal of Mathematics and Mechanics 17, pp. 59–87 (1967).
[10] Bohr, A., Mottelson, B.R. and Ulfbeck, O. "The Principle Underlying Quantum Mechanics." Found. Phys. 34, pp. 405–417 (2004).

"instrumentalism" and "genuine fortuitousness" were adopted by the vast majority of physicists, though most often unwittingly, not everyone renounced trying to understand what was *really* hidden behind the "infamous" quantum probabilities.

4. VON NEUMANN'S REASONING

To understand the nature of the interpretational difficulty that quantum probabilities were unexpectedly posing, we can use again the example of the elastic. Imagine that we are holding the elastic without having yet taken cognizance of its color property. We know the probability that the elastic is black, but we do not know if it really is. On the other hand, we have no doubt that regardless of our knowledge, the color state of the elastic is *perfectly well-defined.*

To describe this situation, the term of *counterfactual definiteness* is usually employed, which simply means that we can speak in a meaningful way even of things which, in fact, we are not observing. For example, we can speak of the position of the Moon although in this precise moment we are not looking at it, and we can talk about the color of the elastic we have in our hand even though we have not yet checked what color it is. We can do this because we have good reasons to believe that the position of the Moon, and the color of the elastic, *are perfectly well-defined properties even when we don't observe them.*

What are these good reasons? Well, considering the example of the elastic, we can say for example that if a colleague would have watched the experiment, s/he could have told us what s/he saw and we, even before looking at the elastic, could have *predicted with certainty* its black color. In the example of the Moon, we can simply mention the fact that every time we looked in the past in the expected direction, according to its orbital motion, we invariably have found it there, faithful to the appointment. In other words, we can predict in advance and *with certainty*, at least in principle, the outcome of our observational process, before executing it.

The amazing thing is that this hypothesis of counterfactual definiteness, perfectly natural and absolutely intuitive, supporting the whole conception of classical realism, seemed to lose validity when it came to consider the properties of microscopic systems, like the atoms and so-called elementary "particles" (which real *particles*, i.e., corpuscles, actually are not). In their case, in fact, it was no longer possible to speak about, say, their position, or velocity, regardless of an actual observation.

To say that an electron had a certain position, or velocity, was meaningful only if the position and velocity were factually observed. Failing that, it was believed that it was only licit to speak of the *tendency* (*availability*, *propensity*, etc.) of these properties in being brought into existence, that is actualized, during a specific observation; a tendency that one could accurately quantify by means of the quantum probabilities, the (apparently non-epistemic) nature of which remained entirely mysterious.

From this unprecedented situation three fundamental questions, intimately linked together, typically arose:

> *(1)* If it is true that the properties of the microscopic entities, as for example their spatial localization, exist only when, during an experiment, they are observed (i.e., measured), what strange mechanism would be responsible for bringing them into existence, i.e., for concretely selecting only one of the many possibilities to which the theory associates the different probabilities?

> *(2)* Since (due to the mentioned impossibility theorems[11])

---

[11] There are theoretical approaches, such as the *de Broglie-Bohm* theory, where the obstacle of impossibility theorems is actually bypassed. This, however, can only happen at the price of having to postulate *ad hoc* the existence of a specific causal field, that would operate at a subquantum level of reality, the random fluctuations of which, once integrated over time, would be at the origin of the ordinary quantum wave function. We will not try here to explain the advantages and disadvantages of the de Broglie-Bohm theory, which presents some serious interpretative difficulties when one tries to describe systems formed by more than a single quantum entity.

quantum probabilities cannot be understood in terms of the observer's lack of knowledge about the state of the system, that is about the properties possessed by the system prior to its observation, what would be then their origin? Or rather: to what kind of lack of knowledge would they refer to?

(3) If, generally speaking, it is not possible to associate *a priori* well-defined values to the different properties of microscopic entities, for instance a specific spatial position, how can we understand the nature of these entities?

Now, considering that laboratory physics' experiments are made by means of *specific measuring apparatus*, corresponding to *macroscopic* entities, of an essentially classical nature, to the first question one might reasonably answer that, regardless of the mechanism which is able to bring into existence (i.e., *actualize*) the *potential* properties of the *microscopic* entities, such a mechanism, necessarily, will have to manifest in the ambit of the *interaction* between the observed system (e.g., an electron) and the specific measuring apparatus used (e.g., a screen detector).

Also, the fact that, once the measurement has been completed, a human observer takes *cognizance* or not of the obtained result, by reading a specific numerical data showed by the measuring device (thus giving rise to a *conscious representation* of the result in her/his mind) should obviously have no relevance in the description of the process.

Yet, many authors believe that the *human consciousness* has a central role in the quantum process of observation-measurement. In the sense that, in ultimate analysis, it would be precisely the conscious representation of the phenomenon in the mind of the investigating scientist that would make it possible, that is real, by actually *selecting* one among the several possible outcomes of the experiment.

If we consider the previous experiment with the elastic, and we do as if the elastic was the equivalent of an electron (and its color the equivalent of the electron's position), then it would be as if the elastic we keep in our hand wouldn't actually possess any specific color, but could only acquire one (in this case, black or white), in a completely unpredictable way (but nevertheless quantifiable in terms of probabilities), in the moment when, by suddenly looking at it, we would become aware of its chromatic property.

But for what reason would some physicists have to consider such bizarre conclusion as convincing? To explain it, we need to reason as did *von Neumann* in the thirties of last century.[12] Von Neumann's argument is roughly as follows. Suppose that it is indeed (as it is reasonable to assume) the measuring instrument (denoted by the letter M) at the origin of the mechanism able to select a specific value of the properties observed in the system in question (denoted by the letter S). To fix ideas, we can think of an *atom* and the measurement of its *spatial position*.

This means that if we consider the interaction between the atom S and the measuring instrument M, at the end of the interactive process S will have acquired, in actual terms, a specific spatial position, among the different positions *a priori* possible; a position that M will be able to explicitly show, for example by indicating on a computer monitor the values of the specific atomic coordinates measured (see Figure 7).

So far so good. Difficulties arise, however, when one considers that in the ambit of quantum theory it is always possible to choose to consider a larger physical system, S', which in addition to S would include in itself also the measuring instrument M, in interaction with S (see Figure 8). And since this larger system S' would also be subject to the quantum laws, its properties would in turn be describable solely in terms of probability!

---

[12] von Neumann, J., *Mathematical Foundations of Quantum Mechanics* (1932); 1996 edition, Beyer, R. T., trans., Princeton Univ. Press.

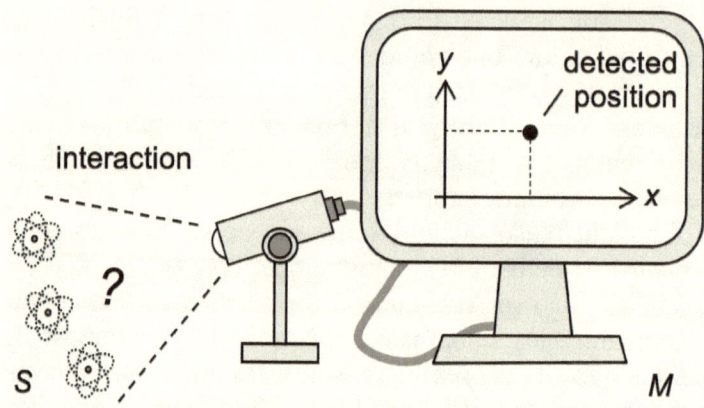

Figure 7. *The measuring instrument M shows a single position for the atom S, among the different positions a priori possible before the observation.*

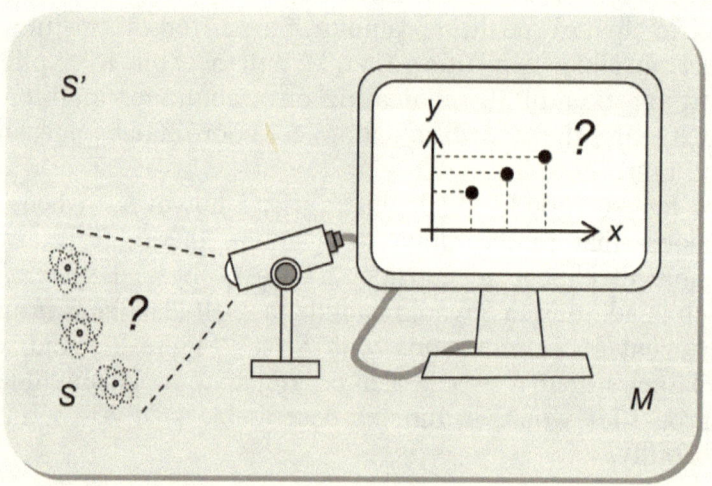

Figure 8. *If the measuring instrument M is described as part of a larger quantum system, S', then, according to quantum theory, it will no longer designate a single spatial position for the atom.*

But in that case, neither the properties of the subsystem S, nor those of the subsystem M, can be described in terms of actualities. And this means in particular that no specific coordinate will be indicated on the monitor of M, in relation to the measurement carried out on S. (More precisely, we should say that in this case there would be different possible screen shots, all potentially present together – in some sense superposed – and each one showing different coordinates for the atom).

One obtains in this way a strange contradiction with respect to the initial hypothesis that it would be the instrument M the cause of the selection process of the observed value. And if one tries to introduce a new measuring instrument M', whose job would be to make a measurement on the system S', assuming again that it is the interaction between S' and M' to be at the origin of the selection process, of course nothing would be solved.

Indeed, one might consider again an even larger system, S'' (see Figure 9), formed by S, M and M', and reapply the probabilistic quantum description to system S'', and so on, in a paradoxical infinite regression.

To escape this impasse, the hypothesis of von Neumann (also revisited in the sixties by *Eugene Wigner*) consisted in affirming that it is the *consciousness* of the investigator – a *non-material* aspect of reality, not subject to quantum laws! – which is the entity capable of activating, in the final analysis, the selection of a specific value for the observed property.

In other words, the simple act of *taking cognizance of the outcome of the experiment* by a conscious mind is what, according to von Neumann, allows a system to move from a situation where the different values of a property are all possible, but no one is concretely actual, to that where only one of these values is brought into existence, that is, actualized in factual terms.

Another way of stating the problem we have just outlined is to make use of the concept of *Heisenberg's cut*. With this concept one refers to a hypothetical "cut" that would allow one to *separate* the observed-system from the observer-system (see Figure 11). Quantum theory gives no indication on where exactly to place such a separation, and this obviously produces the typical contradictions of an argument à la von Neumann, that we have just emphasized.

Figure 9. *The measuring instrument M' is in turn described as part of a larger system, S'', and so on, in a regression without end.*

To avoid these problems, the only possible way out seems to be the one of placing Heisenberg's cut exactly where indicated by the Cartesian dualist ontology, i.e., between the *res extensa* and the *res cogitans*: between the world of the material entities and the one of the observing consciousnesses, whose nature is purely cognitive, that is mental.

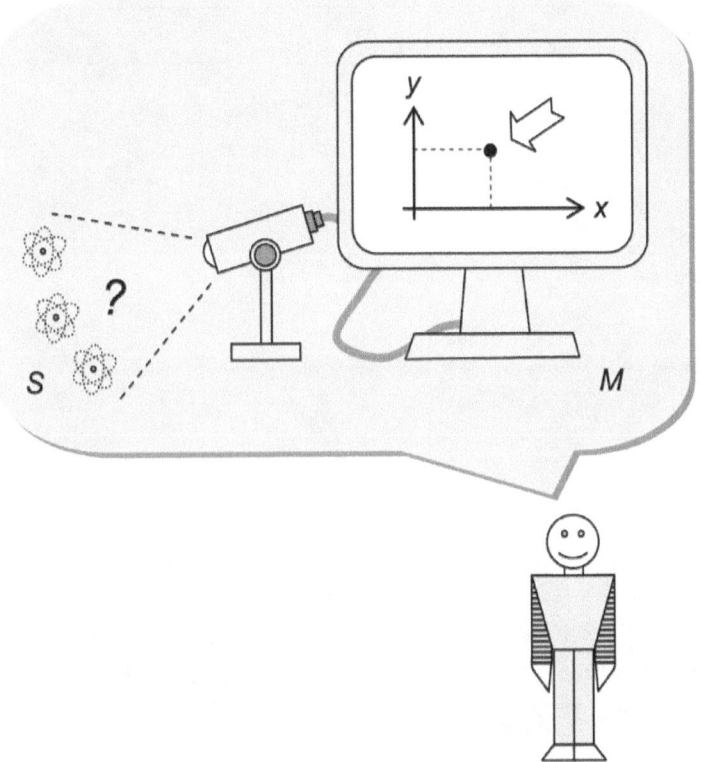

Figure 10. *The conscious representation of the phenomenon by a human observer is what would allow, according to von Neumann's hypothesis, to select a specific outcome in the measurement of the position of the atom.*

Of course, for those who believe that the mind can act directly on matter, as many laboratory experiments in the field of parapsychology appear to suggest, this explanation seems to offer that so hoped opening to found a psychophysical theory of mind-matter interaction, i.e., of the interaction between the cognitive dimension, of a non-material kind, and the dimension of matter-energy.

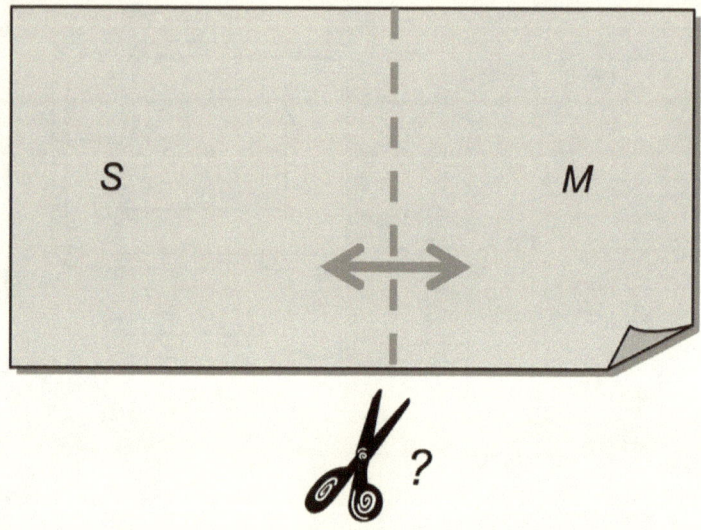

Figure 11. *Quantum theory offers no guidance on where exactly to effectuate the Heisenberg's cut, that is, on how to separate the observed system S from the observer system M.*

Furthermore, if it is true that the immaterial mind can directly influence the matter-energy, then also the fundamental problem of the connection between the body and the spirit could be solved by quantum physics, at least in principle. It would also follow that the existence of the soul, i.e., the ability of the individual to survive the death of her/his physical body, as evidenced by near-death and out of body experiences, would not anymore be an absurdity even for mainstream science, as according to quantum theory the mind would be really able to act independently on the human brain.

For these reasons, it is not uncommon to find in many books

of popular science (but not only), written by authors with a strong orientation towards Eastern or *New-Age* philosophies, statements in which it is argued that physics would have by now demonstrated that a universe without a participatory mind cannot exist, in the sense that it is the mind that would truly shape everything that is perceived, and that all the matter-energy surrounding us would be, in ultimate analysis, a kind of "thought precipitate."

A prime example of this type of position is the famous American documentary of *2004*, titled *What the Bleep Do We Know!?*: a real blockbuster in which it is affirmed, without too much embarrassment, that *according to the laws of quantum mechanics*, human thought would be able to change the nature of our physical reality.

Statements of this kind, however extreme, might appear (partly) justified in the light of the above-mentioned reasoning of von Neumann, who somehow seems to corroborate the thesis that to explain the behavior of microscopic systems it is strictly necessary to consider the element of the human consciousness and the very special nature of its interaction with the physical systems. To quote the words of Wigner:[13]

> *It is the entering of an impression into our consciousness which alters the wave function. [...] It is at this point that consciousness enters the theory unavoidably and unalterably.*

The *wave function* mentioned by Wigner is precisely the mathematical object which describes, in quantum theory, the state of the system, mainly in terms of *potentialities*, and which allows to calculate the *probabilities* associated with the different possible outcomes of an observation.

But however suggestive in metaphysical terms, the hypothesis that the consciousness would be truly responsible for the final concretization of the observational process raises many embarrassing problems, and this explains why the thesis is only

---

[13] Wigner, E. P. *Philosophical Reflections and Syntheses* (annotated by G. G. Emch), Springer (1995).

supported by a narrow minority of physicists.[14]

In fact, if each conscious observer would individually select, in an unpredictable manner, a specific value of the observed physical quantity, how could all the different observers always agree on the different material phenomena they continuously observe in the laboratories? If two observers observe at the same instant the same physical system, considering that each of them would be able to select a different specific outcome, how could their observations always agree?

Certainly, one can always assume that the observation process is activated by the consciousness of the participative observer that first "casts her/his eyes" on the system in question, but is it always possible to make this distinction, for example when two scientists continuously and simultaneously observe the system they are studying?

Leaving aside these perplexing questions, there is in the end a far simpler reason for abandoning the assumption that quantum observation would require the mind of the observer, as emphasized by *Yu* and *Nikolić* in a recent article.[15] In fact, these authors explain, although it is not possible to demonstrate the validity of such hypothesis (since science, as is known, is unable to prove the truth of any hypothesis), it would nevertheless be possible to demonstrate its falsity,[16] that is the falsity of the assumption that the mechanism of *selection of a possibility (SP) necessarily implies* ($\Rightarrow$) a *conscious*

---

[14] See for example: Stapp, H. P. *Mindful Universe: Quantum Mechanics and the Participating Observer.* The Frontiers Collection, Springer, 2nd Edition (2011); Rosennerom, B. and Kuttner, F. *Quantum Enigma: Physics Encounters Consciousness.* Oxford University Press, USA (2008); Menskii, M. B. "Quantum mechanics: new experiments, new applications, and new formulations of old questions." Physics-Uspekhi 43 (6), pp. 585–600 (2000); and references cited therein.

[15] Yu, S. and Nikolić, D. "Quantum mechanics needs no consciousness." Ann. Phys. (Berlin) 523, No. 11, pp. 931–938 (2011).

[16] A so-called scientific assumption is such exactly for the reason that although one cannot prove its truthfulness, one can nevertheless demonstrate its falsity, at least in principle.

representation (*CR*) of the phenomenon in the mind of the experimenter:

$$SP \Rightarrow CR.$$

From a logical point of view, this hypothesis is perfectly equivalent to its *negation* ($\neg$), which is obtained by inverting the terms of the previous relation:[17]

$$\neg CR \Rightarrow \neg SP.$$

This version of the hypothesis, logically equivalent to the previous one, affirms that the lack of conscious representation of the phenomenon in the mind of the experimenter ($\neg CR$) necessarily implies ($\Rightarrow$) the non-actualization of one of the possible outcomes of the observation ($\neg SP$).

Now, many of the already available experimental results (like the so-called "*which-path experiments*," which however we shall not discuss in this essay), if carefully analyzed seem to indicate that the statement "$\neg CR \Rightarrow \neg SP$" is in fact false; so, also the logically equivalent statement "$SP \Rightarrow CR$" would be false, and therefore the alleged link between the human consciousness and the quantum observational process would be in contradiction with the experimental data already in our possession.[18]

---

[17] Let us consider an example: if A corresponds to the proposition "Einstein is a physicist," and B to the proposition "Einstein knows physics," we obviously have that "$A \Rightarrow B$," as the fact that Einstein is a physicist necessarily implies ($\Rightarrow$) that he knows physics (by definition of the concept of "being a physicist"). On the other hand, we can observe that this implication is entirely equivalent to its logical negation, which is obtained by negating the two propositions and by reversing the direction of the implication: "$\neg B \Rightarrow \neg A$." In fact, if "Einstein doesn't know physics" ($\neg B$), it necessarily implies ($\Rightarrow$) that "Einstein is not a physicist" ($\neg A$).

[18] The issue about the possibility of definitively refuting the hypothesis that the mind-matter interaction may be at the origin of the quantum-mechanical collapse of the wave function remains however

That said, and regardless of whether the data today available would or wouldn't have already falsified *von Neumann's psychophysical hypothesis* (or would be able to do so unequivocally), it is important to highlight that the reasons for this strange logico-cognitive impasse reside simply in the belief of many physicists that quantum mechanics would be a complete theory, and that therefore von Neumann's reasoning, although embarrassing, would be in a sense inevitable. On the other hand, the very fact that the theory does not clearly indicate where to put the separation between the observed-system and the observer-system, (the missing Heisenberg's cut) should lead to a certain suspicion about its alleged completeness.

In fact, the incompleteness of quantum theory has already been demonstrated by the Belgian physicist *Diederik Aerts*, more than thirty years ago, although surprisingly this result is still not sufficiently known, or duly considered by most quantum theorists (which often do not even know it). It is indeed possible to show in a mathematically rigorous way that the standard formalism of quantum physics is not at all able, structurally speaking, to describe the simple situation of two *experimentally separate*[19] physical entities, which is typical of many of the macroscopic objects of our everyday life.[20]

This fact is quite severe, if one thinks that to give a proper sense to the distinction between the observed-system and the observer-system, it is obviously necessary to separate them, in

---

controversial; see for example: Acacio de Barros, J. and Oas, G. "Can We Falsify the Consciousness-Causes-Collapse Hypothesis in Quantum Mechanics?" Found. Phys., 47, pp. 1294–1308 (2017).

[19] Two entities are said to be *separate*, in *experimental* terms, if the execution of an observational experiment on the first entity does not affect the outcome of a (simultaneous or sequential) observational experiment performed on the second one, and vice versa. Note that the concept of experimental separation does not necessarily imply that the two systems in question would be non-interacting.

[20] Aerts, D., "Description of many physical entities without the paradoxes encountered in quantum mechanics." Found. Phys., 12, pp. 1131–1170 (1982); "The missing element of reality in the description of quantum mechanics of the EPR paradox situation." Helvetica Physica Acta, 57, pp. 421–428 (1984).

the sense that in an observational process the observation system M should be, typically, initially separated from the observed system S, then the two systems should connect in some way, to allow the measurement per se to take place, and at the full completion of it separate again. But such a process of separation-union-separation is not at all describable within the restricted framework of conventional quantum physics, hence the need for the extra-systemic concept of an immaterial consciousness, in order to try to solve the issue.

On the other hand, there exist theoretical approaches to the description of physical systems that are much more general than quantum mechanics, able to describe both purely quantum systems, whose parts cannot be analyzed separately, and purely classical systems, which instead can be separated, with the addition of a new class of systems of an intermediate nature, called *quantum-like*, which are true hybrids, halfway between classical systems and quantum systems.

These are approaches that even prior to considering the specifications of the microscopic world, they try to identify and clearly describe what are the "rules of the game" when a physicist studies in all generality (both theoretically and experimentally) a material system, be it macroscopic, microscopic, or mesoscopic.

Obviously, we are here at the boundary between physics and philosophy of knowledge, a delicate area conceptually speaking, where not all researchers feel at ease. But this is the territory where it is necessary advancing, if one wants to penetrate some of the mysteries and oddities of the reality of the microworld. Indeed, by studying the foundations of physical theories in a broad sense, one can realize that some of the peculiarities of the microworld are in fact already present in our interaction with the conventional macroscopic entities, if only we learn to observe the content of these interactions/observations with the due discernment and from the right perspective.

This, at least, is what has emerged from the findings of the so-called *Geneva-Brussel school on the foundations of physics*, which originated in the pioneering work of *Josef-*

Maria Jauch[21] and *Constantin Piron*,[22] in Geneva, and found its maturation in the fundamental works of *Diederik Aerts* and his group in Brussels.[23]

Contrary to what has been done in the past, during the development of orthodox quantum theory, instead of first deriving a formal mathematical structure, and only subsequently trying to see what could be its physical interpretation, the founders of this school "rolled up their sleeves" and went back to that more natural method, which consists in trying to initially identify what the relevant physical concepts are, defining and clarifying them on a solid *operational* and *realistic* basis, and only then use them to build a mathematical theory of the physical reality, which then will have greater chances to be entirely meaningful and intelligible.

Following this more satisfying approach, researchers of the Geneva-Brussel school (and more particularly Aerts) managed in the years to derive a very effective conceptual and mathematical language, called the *creation-discovery view*,[24] able to describe the different dynamics of the entities that populate our reality with a high level of generality and universality.

In this way, it was possible to elucidate many of the conceptual oddities and ambiguities present in the different interpretations of quantum physics, developing an approach (today still under study and perfection) with which it becomes

---

[21] Jauch, J.-M. *Foundations of Quantum Mechanics*, Addison-Wesley Publishing Company, Reading, Massachusetts (1968).

[22] Piron, C. *Foundations of quantum physics*. Massachusetts: W. A. Benjamin (1976); *Mécanique quantique: Bases et applications*. Presses polytechniques et universitaires romandes, Lausanne, Switzerland (1990).

[23] Although this school is still indicated as the *school of Geneva-Brussel*, it is currently only active in Belgium, especially in the ambit of the *CLEA – Center Leo Apostel for Interdisciplinary Studies*, at the *Vrije Universiteit*.

[24] Aerts, D. "The entity and modern physics: the creation-discovery view of reality". In: *Interpreting Bodies: Classical and Quantum Objects in Modern Physics*. Ed. Castellani, E. Princeton Unversity Press, Princeton, pp. 223–257 (1998).

possible to study the behavior of entities both physical and non-physical (such as cultural entities, signs and symbols, concepts, minds, etc.).

Of course, we will not enter here into the merits of the sophisticated mathematical and conceptual language of this school of thought, which is very rich and elaborate. We will limit ourselves here to follow one of the traditions of this school – especially with respect to the work of Aerts – which is the one of inventing and analyzing very simple (but not for this less significant) *macroscopic models*, able to incorporate all of the strangeness of the behavior of the microscopic physical systems. This strangeness, however, as it will fully reveal before our eyes, will definitely appear much less mysterious than expected.

More precisely, what we propose to do in the following pages is to continue our observational experiment with the elastics and show how a seemingly simple and conventional macroscopic system as this is in fact able to offer satisfactory answers to the three fundamental questions we have mentioned above.

5. Quantum Elastic Bands

If previously we were interested in the color (white or black) of the elastics, with the aim of emphasizing the nature of *classical probabilities*, and of the properties associated to them, whose actuality is independent of our observation, we now want to consider a quite different class of properties, which will allow us to reveal the deep nature of *quantum probabilities*, and of the mysterious *selection mechanism* which guarantees that in an observation one moves from the level of abstract possibilities (the different possible outcomes before the experiment) to that of a concrete actuality (the specific outcome factually observed in the laboratory, following the observational experiment).

Before defining these properties, it is important to open a brief parenthesis and ask the following question:

> *How do we generally proceed in order to attribute properties to the different entities belonging to our physical reality?*

That is, why do we attach, for instance to an elastic band, the property of "having a color"? The question may seem strange, but as we will understand the reasons for those attributions have some relevance in our analysis. One possible response, that we already mentioned, is that since in our past interactions with the elastics we have found that these always had a specific color, it is quite natural to assume that the elastic we held in our hand earlier also possessed its own specific color, and therefore the question we addressed ourselves on its possible black color was perfectly licit and appropriate in relation to this kind of object.

The same reasoning can of course be applied to many other properties related to physical entities, such as that of "having a given position in space." Indeed, we have always found, in our many interactions with the material objects, that they have a specific location (although this location can obviously change over time), and it is therefore perfectly legitimate to question ourselves about the spatial position of a particular entity, such as the Moon, at a given time.

To simplify the discussion, we can say that we humans, in the course of our biological evolution, have gained some experience about the (macroscopic) material entities that form our ordinary reality, and have found that a number of properties can be sensibly and stably applied to them. From this experience, like it or not, a *prejudice* resulted, which is the one of believing that we can generalize without problems the results of our observations, considering for instance that since the concept of position can be suitably applied to the macroscopic objects of our everyday life, the same should hold true also for those microscopic "objects" we can detect through appropriate observational instruments.

But here we must be careful not to commit what logicians call a mistake of *hasty generalization*, consisting in reaching a general conclusion on the basis of information obtained on a sample which is not necessarily representative. Obviously, there is no logical basis for such an inference, as we have never directly interacted with the microscopic entities, like atoms, and this for one simple reason: microscopic entities, exactly because they are such, are invisible to our *ordinary* perceptual instruments. The aforementioned prejudice, as we shall see, is at the basis of our lack of understanding about the true nature of microscopic entities, as in fact it is a *false prejudice*.

The reason for this short parenthesis is to highlight the following:

> As it is unquestionably a *non-ordinary procedure* to attribute *ordinary properties* to *non-ordinary entities*, such as electrons, in the same way – although for opposite reasons – it is as much a non-ordinary procedure to attribute *non-ordinary properties* to *ordinary entities*, such as an elastic band.

What we are trying to highlight is that what determines the *classical* or *quantum* nature of a physical entity is not so much the fact that it is *macroscopic* or *microscopic*, but the nature of the questions we ask in *operational* terms in relation to it, i.e., in relation to the properties that we believe we can attach to it, and therefore observe.

The important point to understand is that the ordinary or non-ordinary (i.e., classical or non-classical) character of a question does not depend on the question itself, but on its specific relation with the entity to which the question is addressed. In fact, if asking a question about the position of the Moon is perfectly ordinary, we cannot say the same if the same question is asked in relation to an electron.

So, a good strategy for understanding the nature of the quantum reality is to ask genuinely non-ordinary questions in relation to ordinary macroscopic entities, and then see how they react to these strange questions of ours. In fact, the advantage here is that since a macroscopic entity is constantly before our eyes, it becomes possible to understand what really happens when, during an observation, we implement in practical (experimental) terms our non-ordinary questions.

As we shall see, this will allow us to solve the mystery of the origin of quantum probabilities and of the counterintuitive "observer effect," on the basis of the so-called *hidden-measurement approach*, originally proposed by Diederik Aerts.[25]

Consider then a simple elastic band. We shall not anymore be interested in its color, but will try to determine if the elastic

---

[25] The *hidden-measurement approach* was deduced by its author starting from an accurate analysis of specific macroscopic *quantum machines*, whose behavior was surprisingly able to imitate the one of microscopic systems. In this booklet we will not enter into the details of Aerts' specific quantum machines, since they would require, to be understood, a specific knowledge of the microscopic quantum systems they refer to. Nevertheless, the analysis we will present of some *non-ordinary* properties of a simple elastic band, already contains in itself those conceptual ingredients that will allow us to elucidate the possible origin of quantum probabilities in the microscopic systems.

possesses or does not possess a particular property, which we will call *left-handedness*.

Figure 12. *A simple elastic band, in the present case of black color.*

This property is of the *non-ordinary* kind for an elastic band. In fact, we are not used, in our ordinary interactions with these entities to ask ourselves about their left-handedness. Actually, we don't even know what left-handedness could mean in relation to an elastic band!

Therefore, before we can observe the left-handedness of an elastic band, we must define what it would mean to actually observe it. It must be said that the procedure consisting in defining a property by means of a precise description of the operations to be carried out in order to observe it is quite natural in physics, and is called *operationalism*. So, to define what the left-handedness of an elastic band is, we simply have to explain how to observe it in practical terms. The observational procedure is very simple and is as follows:

Left-handedness (respectively, *right-handedness*) *observational protocol*: Grab the two ends of the elastic with both hands (see Figure 13), then stretch it strongly and abruptly, so it breaks. At this point, observe the two fragments dangling from your hands. If the longer one remains in the *left* hand (see Figure 14), then by definition the elastic is left-handed; otherwise it possesses the inverse property of right-handedness (see Figure 15).

left-handed or right-handed?

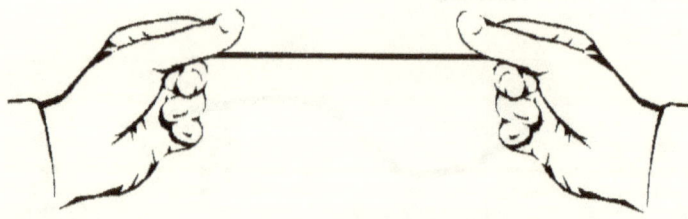

Figure 13. *The elastic band is grabbed by its two ends by the hands of the experimental scientist, before s/he proceeds to the pull, which will determine the outcome of the observational process.*

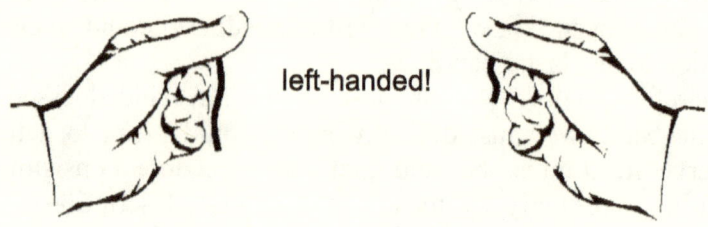

left-handed!

Figure 14. *The observation of the left-handedness is successful when the longer fragment remains in the left hand.*

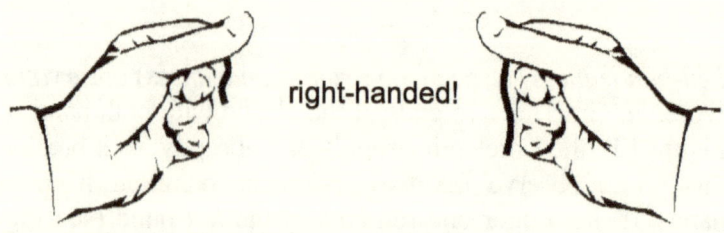

right-handed!

Figure 15. *The observation of the right-handedness is successful when the longer fragment remains in the right hand.*

Of course, it may happen that an elastic band, because of its previous interactions with other physical entities, no longer forms a whole, but is composed of multiple spatially separated fragments. In other words, it could happen that an elastic is already broken into several pieces. We can nevertheless continue to speak in a sensible way of the left-handedness (or right-handedness) property, simply by adding to the above-mentioned observational procedure the condition that if the elastic is already broken, the experiment has to be conducted using the longer fragment.

One could argue that it is meaningless to speak of an elastic when it is broken, but the definition of what is or is not an elastic is clearly conventional, and therefore in our way of thinking about these entities we are totally free to include in their definition the fact that they can either form a cohesive whole, or be composed of several separate pieces, without for this losing their primary elastic-identity.

Figure 16. *An elastic band, according to our definition, remains such even if composed of several separate fragments (in this case two), that is, even if it is broken.*

One could also argue that in some rare cases it may happen that an elastic band is found to be *ambidextrous*, in the sense that the fragments in the two hands, following the pull, are of equal length (within the limits of measurement accuracy). To avoid unnecessary complications (which would add nothing more to our discussion) we will assume in the following that it is always possible to determine which of the two fragments is in fact the longest.

Having clarified how the property of left-handedness (or right-handedness) has to be precisely understood, in relation to an elastic band, imagine now that you are holding in your hands an elastic and that you ask yourself the following question:

| *Is the elastic band left-handed?*

As for the previous question on the black color, also this question admits of course, in principle, only two responses: *yes*, or *no*. And as for the previous question, you are not in the condition of answering until you have completed the observational experiment.

In the previous experiment, relative to the black color, the inability in answering the question resulted from the fact that you hadn't yet *taken knowledge* of which was the elastic drawn from the box, as you hadn't yet looked at it. So, without first completing the experiment, by looking directly at the elastic, you could only say that the answer to the question was "yes" with probability of *50%*. But then, once you have directly looked at the elastic and determined that its color was actually black, to any further question about its black color you could have answered "yes" with absolute certainty, i.e., with probability equal to *100%*.

Let us now see what changes with the left-handedness. Again, we can answer the question without any major problems in probabilistic terms. In fact, by getting a sufficient number of elastic bands, all identical to what we have in our hands, we could have performed in advance the observational experiment of the left-handedness on each of them, and determine the percentage of positive outcomes. Suppose that this percentage is exactly *50%*, as is easy to deduce if we consider that the observational procedure does not favor in any way the left with respect to the right.

Thus, using a simple theoretical reasoning, or our previous statistical study, when we hold the elastic in our hands we can state that it is left-handed with a probability of *50%* (and consequently right-handed with equal probability), just as we were previously able to state that it was black with a probability of *50%* (and consequently white with the same probability).

But now think. When the question was for the color, the probability of *50%* was due to the fact that we hadn't yet looked at the elastic. But in the case of left-handedness, we're not looking away, or closing our eyes: we can perfectly see the elastic stretched between our two hands!

So, the question arises: since, in spite of the fact that we are looking at the elastic band, we are not able to answer with certainty to the question about its left-handedness: *what is it that we are not "seeing"*? Or rather, is there perhaps something that we could know about the elastic, for example some of its *hidden properties*, which would allow us to predict with certainty the outcome of the left-handedness' observational experiment? Is there a way to study the elastic in question, perhaps obtaining information concerning its manufacturing method, the quality and precise characteristics of the rubber used to produce it, etc., which would allow us to answer the question in non-probabilistic terms?

The answer to this question, as it is easy to convince oneself, is negative. Even with a complete knowledge of all the characteristics of the elastic, up to the level of its molecular structure, there would be no way for us to establish *a priori* whether it is a left-handed or right-handed elastic. This is so for one simple reason: the property of being left-handed (or right-handed) does not already exist (it is not yet actual, but only potential) for that elastic!

> *The left-handedness (or right-handedness) of an elastic band is created (i.e., actualized) by the process itself of its observation, and this process is in no way under the control of the observer!*

And since it is not under her/his control, s/he has no chance to predict its outcome in advance.

6. THE SOLUTION OF (PART OF) THE ENIGMA

Thanks to the example of the left-handedness of the elastic bands, we are now in a position to solve part of the enigma and give a conceptually satisfactory answer to the first two of the three questions that we have previously addressed.

If you remember, the first question concerned the mechanism able to select a specific result among the different possible outcomes, otherwise described in probabilistic terms. As in the case of the color, which could be black or white, also in this case the different possible outcomes of the process of observation are only two: the elastic is left-handed (the longer fragment is in the left hand) or right-handed (the longer fragment is in the right hand).

What determines the outcome? Obviously, the outcome is determined by the exact point x where the elastic will break, when we pull it strongly with our two hands. If x is in the half of the elastic close to the right hand, the outcome of the observation will be the left-handedness. If on the contrary x is located in the half of the elastic close to the left hand, the outcome will be the right-handedness (see Figure 17).

One way to describe the process is to say that to every possible breaking point x, there corresponds a specific way (or more precisely, a number of specific *equivalent* ways) to stretch the elastic band so as to produce its breaking exactly at that point. Each class of equivalent ways to stretch the elastic corresponds to a specific interaction I_x (or more precisely, to a specific class of equivalent interactions) between the measuring apparatus constituted by the two hands of the observer and the physical system constituted by the elastic.

What is important to understand is that because of a number of fluctuating factors, such as the subtle vibrations of the hands while they are pulling, the specific orientation of the elastic when it is grasped, the pressure exerted by the fingers, the rapidity with which the elastic is stretched, and so on, it is truly impossible for the experimenter to control which exact interaction I_x will be actually selected, among all the possible ones, and thus to predetermine at which point x the elastic will eventually break.

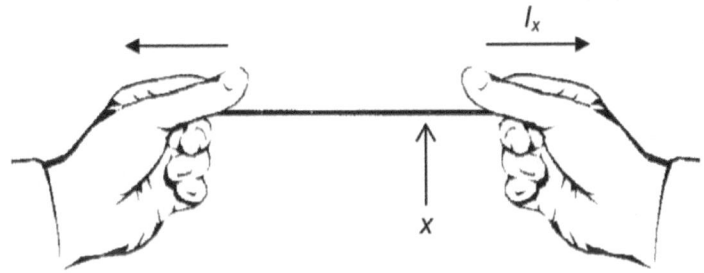

Figure 17. *To every interaction Ix (which in fact corresponds to an entire class of equivalent ways of pulling the elastic) corresponds a specific breaking point x of the elastic. In the present case, the interaction Ix which is (unconsciously) selected by the hands of the experimenter will determine the left-handedness outcome.*

Before the execution of the observational experiment, all breaking points are *a priori* possible. Therefore, they are all *potential breaking points*, not yet actual, as is only potential the left-handedness (or right-handedness) property of the elastic. But when the elastic is grasped and strongly stretched, a real *symmetry breaking* process occurs, in the sense that a specific interaction I_x is accidentally selected by the observer, which in turn will cause the rupture of the elastic in a specific point x.[26]

---

[26] The concept of *symmetry* is here to be understood in the sense that all points x (or all classes of interaction I_x associated with them) are *a priori* equivalent (in the sense of being *equiprobable*) from the point of view of the breaking of the elastic (being the latter by hypothesis uniform). One way to make this symmetry fully manifest is to

Accordingly, the property of left-handedness will be either confirmed or disconfirmed, i.e., its observation will be either successful or unsuccessful.

So, we are now able to explain how the selection of a specific outcome occurs, among the many *a priori* possible ones, when these are describable only in terms of probabilities, without, however, that these probabilities would be attributable to a lack of knowledge of the observer concerning the properties of the observed system (as is the case instead for classical probabilities).

The selection mechanism is the typical one of a *symmetry breaking* (what is actualized breaks the symmetry of what was potential), which occurs as a consequence of the inevitable presence of uncontrollable and unpredictable fluctuations in the process of observation. Because of these fluctuations, the observer is not in a position to control which specific interaction I_x, between the observed system and the observation instrument, will be actually selected, determining in this way one only among the possible outcomes of the measurement (which in our simplified example are only two).

Once the mechanism has been identified, we are also able to elucidate the mystery of the origin of quantum probabilities, i.e., the nature of those probabilities that cannot be explained in terms of lack of knowledge of the observer concerning the properties possessed by the system before its observation.

As can be seen from the experiment with the elastic band, the probabilities associated with the observation of the left-

consider, instead of an elastic band with two extremities, a ring-shaped one, like the typical office or school elastics (in this case, when making the observation of the left-handedness, the elastic will break in two points, rather than just in one). With a ring-shaped elastic the symmetry about which is here the question can be visualized as a simple *rotational symmetry*. In fact, it is possible to arbitrarily rotate the elastic prior to the observation, without for this that the effect of the rotation will affect the probabilities of its possible outcomes. This *rotational invariance* is therefore an expression of a symmetry of the elastic, which in turn is an expression of its uniformity.

handedness are also *epistemic* in nature – that is, associated to a state of ignorance of the observer – but the lack of knowledge to which they refer is of a different nature than in so-called classical probabilities. In fact, this lack of knowledge has its origin in the *lack of control* by the experimenter about the details of the execution of its own process of observation; a lack of control which, as it results in a lack of knowledge regarding the final outcome of the experiment, can nevertheless be quantified in probabilistic terms.

So, if we want to talk about *hidden variables*, these are to be associated not with the observed system, that is, to its state, but with the process of observation, i.e., with the measuring act itself. That's why the explanation of the origin of quantum probabilities suggested by models such as the one of the elastic, has been named by its discoverer *hidden-measurement approach*.

Certainly, one could object that the ultra-simplified example of the left-handedness experiment of an elastic is just a metaphor, but it is not so. In fact, one can exploit the crucial ingredient of the "hidden observations" (in the sense of "hidden measurement interactions") to build, in all generality, a probabilistic theory of a non-classical kind, and show that in the limit of a complete control of the act of observation by the observer, it reproduces a classical probabilistic theory (obeying the classical axioms of Kolmogorov), while in the opposite limit of a complete ignorance about the nature of the selected interaction I_x, it yields the typical (non-kolmogorovian) probabilistic structure of quantum mechanics.

Furthermore, as already mentioned, it is possible to highlight intermediate situations, halfway between the classical one, of a full control of the interaction, and the quantum one, of a complete absence of control of the interaction. These intermediate, *quantum-like*, situations, describe observational regimes which are much more general, impossible to describe within the restricted formalism of conventional quantum (or classical) mechanics, and undoubtedly more suitable to account for the different possible articulations between the countless entities that populate our reality.

7. EPHEMERALITY

Before moving to the third question, the one concerning the nature of the microscopic quantum entities, which by the way is the question that hides the real mystery, it is important to spend some more words on the characterization of a quantum observational process.

What we have so far emphasized, by means of the paradigmatic example of the observation of the left-handedness, is that a quantum observation contains a mechanism of *symmetry breaking*, which selects in an uncontrollable (and therefore unknown) way a specific interaction between the observed entity and the instrument of observation. In other words, we can say that each single observational process of a quantum property consists in fact of an entire collection of possible observational processes (of possible measurements), and that unbeknownst to the observer only one of these processes will in the end be selected, determining in this way a specific outcome. In technical language, this first ingredient characterizing the observations of a quantum kind is called *product observation*, or *product test*.

The second ingredient that emerges from our analysis of the elastic is the *creation aspect*. It is evident that the property of left-handedness is only a potential property before the act of observation, which may become actual only after it, depending on the type of interaction that will take place between the hands of the experimenter and the elastic. To put it more simply, such a property is not ordinarily attributable (in the *actual* sense of the word) to an elastic band when it is in its standard condition. In fact, for an elastic that of the left-handedness is an exceptional property, that it is necessary to specifically create by means of a non-ordinary procedure, associated with an as much non-ordinary question. And it is the procedure by which the question about the left-handedness is implemented in practical terms (i.e., operationally), which is responsible for the possible creation of the same left-handedness.

There is also a third ingredient, which in a sense is a logical consequence of the aforementioned "creation aspect" in relation to non-ordinary properties, i.e., to properties not ordinarily possessed by an entity. The third ingredient is the one of the *ephemeral* character of the observations of the quantum kind. By the term "ephemeral" herein is meant the fact that the property that is (possibly) *brought into existence*, that is, *actualized*, by the act of observation, becomes again *potential* at the end of it, that is, it is ultimately destroyed (de-actualized).

To understand this better, consider again the experiment of the left-handedness. Suppose you have created the left-handedness after having broken the elastic with your two hands. This condition of *manifest left-handedness* remains such only for as long as the specific *relation* between the hands of the experimenter and the two pieces of elastic is maintained, that is, for as long as the left hand holds the longer fragment and the right hand the shorter one (see Figure 14).

In fact, it is in the manifestation of such a *relational property* that the left-handedness property of the elastic becomes *actual*. On the other hand, as soon as the experimenter-observer drops the fragments of the elastic (see Figure 18), the latter immediately loses its left-handedness relational property, which again becomes a genuinely potential property, no longer manifest.

Let us remember that the observational procedure of left-handedness, as we have defined it, also applies to an elastic band consisting of several separate fragments. So, once the relationship with the hands of the experimenter is lost, to observe another time the left-handedness (or right-handedness), s/he will have to pick up the longer fragment, break it again according to the procedure, and observe whether the longer fragment is once again in the left (or right) hand.

To cut a long story short, at the *complete* end of the observation (which in this case means having dropped the elastic fragments), left-handedness disappears from our sight, that is, from our *observational space*, and to observe it again it is necessary to create it again, using a new observational process, whose nature is however entirely unpredictable.

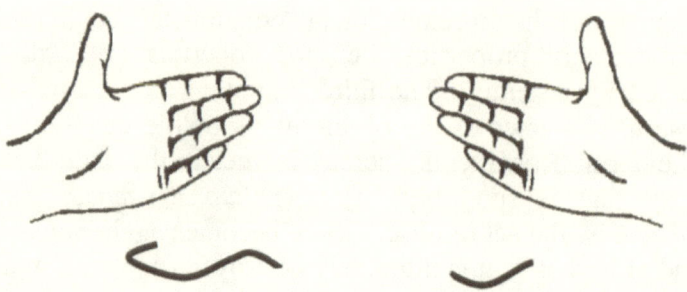

Figure 18. *As soon as the hands of the experimenter-observer drop the pieces of elastic, the property of left-handedness, as it was created, is destroyed (de-actualized).*

Now, this *ephemeral* character of the left-handedness property, we find it as such also in the observation of the properties of microscopic entities, like for example the *spatial localization* of an electron. This can be evidenced (although in a non-deterministic way) through the interaction of the electron with an appropriate measuring instrument (e.g., a screen detector), by establishing with the instrument a specific *relation*.

But as indicated by the same formalism of quantum theory, the outcome of a subsequent observation of the electron's localization (separated by a finite time interval from the preceding one) will again be non-predeterminable, demonstrating that the property of possessing a specific localization is not acquired in a stable manner by the latter, but in general needs to be recreated at each subsequent observation.

To summarize, we can conclude, thanks to our analysis of the paradigmatic example of the left-handedness, that the following three properties characterize a quantum observational process, and explain the origin, in quantum mechanics, of the mysterious "observer effect":[27]

> (a) Presence of a *symmetry breaking* mechanism selecting one among several possible interactions between the observed entity and the observing instrument, in a way *not directly controllable by the observer*.

> (b) Presence of an *invasive process of creation* of the observed property, which doesn't exist (is not actual) prior to the observation.

> (c) *Ephemeral* nature of the created property, which becomes once more potential at the end of the observation.[28]

---

[27] Actually, one can show that the third property is not really independent from the first two, but in fact follows from them. See: Sassoli de Bianchi, M. "God may not play dice, but human observers surely do." Foundations of Science 20, pp. 77–105 (2015).

[28] This is so because of the *purely relational nature* of the property in question, which necessitates the creation of a specific relation, with a specific observational system, in order to be actualized.

8. THE (TRUE) ROLE OF THE OBSERVER

Of course, much could (and should) be added regarding the nature of a quantum observational process, but within the limits of space and the expositional logic of this booklet, we cannot go into more details in this analysis.

The careful reader will have certainly noticed that the third question that we have addressed, which actually contains the real mystery about the nature of the microscopic entities inhabiting our physical world, is still open. Indeed, if it is true that we have no means of attributing in a stable way, to the entities of the microworld, those properties that are instead intrinsic properties of the macroscopic entities – such as having a position, or a velocity – how can we hope to understand their nature?

Before answering this fundamental question, and lift a corner of the veil that shrouds the entities of the microworld, it is important, on the basis of our previous analysis, to take stock of the situation about the actual role of the observer. What we can say, considering the experiment of the left-handedness as an archetype of a quantum process of observation-measurement (during which a potential property is brought into actual existence in a totally unpredictable way, as a consequence of the invasive action of the experimenter over the observed system) is the following:

> It is absolutely unnecessary, in order to explain the quantum probabilities and the selection process they subtend, to bring into play the mind of the experimenter. *There is no quantum observer effect of a psychophysical nature,* that is, an effect that would be caused by the action of the immaterial mind of the observer on the observed material system, but rather an *effect of the observational instrument* used by the observer, which may be either a machine, employed to extend her/his

observational possibilities, or her/his own body, as in the case of the left-handedness' experiment with the elastics.

To avoid possible misunderstandings, it should be noted that although quantum mechanics, according to the analysis proposed here, doesn't allow to infer about a direct action of the psyche of the observer on the observed systems, this of course does not mean that when a human observer mentally focuses on certain aspects of reality, her/his attention (which could for example be aimed at obtaining a specific result) could not have some influence on these aspects. In other words:

The fact that a direct action of the consciousness on the matter-energy is not a necessary ingredient to explain observation in quantum physics does not necessarily mean that such an action would be impossible.

This kind of conclusion would constitute a clear error of reasoning, as it would be wrong to believe (as some parapsychologists often do) that an experimental confirmation of the PK-effect (psychokinesis) in relation to microscopic systems would necessarily imply that the reasoning of von Neumann as regards a psychophysical interpretation of the quantum observational process is to be considered correct.[29]

As already mentioned, there is a considerable body of data supporting the hypothesis of an interaction between the mind and the matter-energy,[30] although the interpretation of these data is still controversial to this day. It is therefore not possible to rule out the existence of mechanisms, yet to be elucidated, that would explain the possibility of such a "subtle action." But

---

[29] Radin, D. et al. "Consciousness and the double-slit interference pattern: Six experiments." Physics Essays 25, pp. 157–171 (2012); Sassoli de Bianchi, M. "Quantum measurements are physical processes. Comment on 'Consciousness and the double-slit interference pattern: Six experiments', By Dean Radin et al. [Physics Essays 25, 157 (2012)]". Physics Essays 26, pp. 15–20 (2013).

[30] See for example: Krippner, S. and Friedman, H. L., Editors. *Debating Psychic Experience: Human Potential or Human Illusion?* Praeger (2010); and the references cited therein.

the point of this discussion is that these mechanisms, if they exist, would not be amenable to modern quantum theory, not directly at least.

Instead, what quantum physics has undoubtedly highlighted is the existence of a number of prejudices about our understanding of the very concept of observation.

> *To observe is not (always) a neutral activity, but an activity that can involve both elements of discovery and of creation (or destruction).*

Typically, we are used to think of observation only in terms of *discovery*, i.e., as an instrument that allows us to discover those things that already exist "out there." On the other hand, although the majority of human beings understand well what a creative (or destructive) process is, it is usually not considered in relation to the process of observation.

Generally, we agree in thinking that a process of observation may possibly involve a certain element of *disturbance*, as is clear that the *sine qua non* condition for an observation is that the observer system interacts in some way with the observed system. On the other hand, there are systems which emit spontaneously information to the outside, as is the case for example of a light source, and therefore there are observational situations where the disturbance is by definition zero (one speaks then of a perfectly *non-invasive* observation).

On the other hand, according to the *classical prejudice*, it should always be possible to make the disturbance arbitrarily small (using increasingly sensitive and less invasive observational instruments) and consequently, in ultimate analysis, reduce each observational process to a process of pure discovery. The example of the observation of the left-handedness of an elastic band, however, clearly disproves such a belief.

Of course, we must agree here on the definition we want to attribute to the term "observation." We can either extend this concept, by incorporating in the act of observation also the possibility of creation or destruction of the observed properties, or describe these possibilities by different terms. But if we do so, we risk ending up in awkward semantic situations, when describing certain processes.

As an example, consider a simple match, and ask what it means to *observe its ignitability*. We hope it is clear to everybody that there is only one way to observe the ignitability of a match: to strike it and see if it lights and burns. It is also clear, however, that the observation the ignitability of a match in turn *destroys* the property, as is known that burnt-out matches are no longer ignitable matches! (See Figure 19).

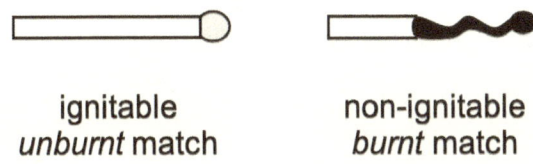

ignitable
unburnt match

non-ignitable
burnt match

Figure 19. *It is important to distinguish unburnt ignitable matches from burnt non-ignitable ones, as they do not possess the same properties.*[31]

In ultimate analysis, our observations are nothing but *tests* we perform, often without our knowledge, about the properties that are possessed (or can be possessed) by the different entities populating our reality. When we look at the trees, walking in a forest, we do nothing but test their presence, their spatial position, their color, etc. Obviously, in this case the test subtended by our observational process is totally non-invasive (since the trees, in their usual state, spontaneously reflect the sunlight), and the observed properties are intrinsic to the object, in the sense of being observable in the same way (or in

---

[31] The author still remembers with surprise when, at the Geneva's School of Physics, he heard for the first time Constantin Piron, while gesticulating at the blackboard, solemnly warning his students of quantum mechanics about the danger of confusing *breakable chalks* with *broken chalks*!

a perspectively equivalent way) by any other observer walking in the forest.

But in other ambits, the observational process can easily create those same properties that are meant to be observed. The point is: *are we aware of this, or does this happen without our knowledge?*

In the case of the observation of quantum entities, the creation aspect has certainly been revealed to us without us being initially aware of it. That's why we were so surprised by the results of our measurements, when we became interested in microscopic entities, and that's why there are many physicists who have not spared their cognitive efforts to bring back the quantum observations to processes of pure discovery, as is the case for example of *Luis de Broglie*'s *pilot wave* approach, subsequently reworked by *David Bohm*, or of the *quantum decoherence* approach.[32]

---

[32] We shall not analyze in this writing, for obvious reasons of space and purpose, the advantages and disadvantages of these and other approaches, in relation to a possible solution of the problem of quantum observation (also called quantum measurement problem). For *Bohm*'s interpretation, we refer the interested reader to the discussion in Aerts, D. *"The Stuff the World is Made of: Physics and Reality."* In: *The White Book of 'Einstein Meets Magritte'.* Edited by: Diederik Aerts, Jan Broekaert and Ernest Mathijs, Kluwer Academic Publishers, Dordrecht, pp. 129–183 (1999). As regards instead the *decoherence* interpretation (today quite widespread among physicists), we recall only that it does not really solve the problem of the origin of the selection mechanism producing the transition from probabilities to actualities, but merely shifts it from the system to its environment. Among the best known approaches, we can also remember the exotic *many-worlds interpretation*, suggested by *Hugh Everett III*, the *transactional interpretation* of *John Cramer*, which introduces the existence of processes able to propagate backward in time, the *relational quantum mechanics* of *Carlo Rovelli*, and the *objective collapse* theories, such as the one developed by the Italian physicists *Giancarlo Ghirardi, Alberto Rimini* and *Tullio Weber*.

9. CREATION AND UNPREDICTABILITY

The example of the observation of the ignitability of a match is obviously of a destructive kind, as it is destructive the observation of the *solidness* of a car, in a typical *crash-test*. But of course, we can easily imagine observational processes that are able to literally (and permanently) create the observed property. Let us take an example.

Consider a small solid object, of whatever shape, made of a non-elastic material, and suppose we want to observe its *incompressibility*, which for the purposes of the present discussion we will define as follows:

> *Incompressibility observational protocol*: Take the solid object and subject it to the action of a press capable of exerting a pressure of *10,000 pascal*. If following the action of the press the volume of the object is not reduced by more than *1%*, then it is by definition *incompressible*; otherwise it possesses the inverse property of *compressibility*.

Of course, when we carry out the observation, that is, when we submit the object in question to the action of the press, the result of the action can be either positive (the volume is reduced by a percentage less than *1%*) or negative (the volume is reduced by a percentage more than *1%*), depending on the type of material and the shape of the object.

In the case in which the outcome of the observation is positive, we can conclude that we have actually observed the *incompressibility* of the object. However, we certainly cannot claim that we have created the property of incompressibility through the process of observation, as this property was clearly already possessed by the object even before its observation.

What can we say instead of the situation where the volume reduction is greater than *1%*? Obviously, since the outcome of the observation is negative, we have failed in this case to observe the incompressibility of the object, and have instead confirmed its *compressibility* (see Figure 20).

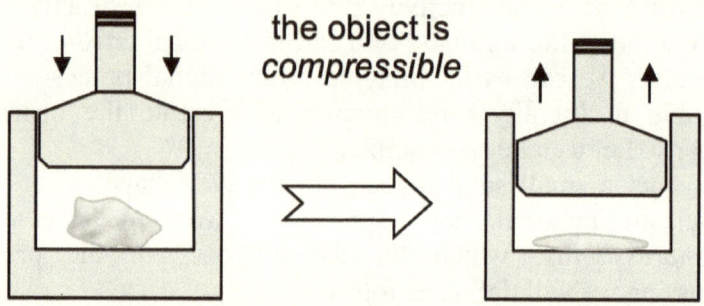

Figure 20. *The negative result of the observation of incompressibility (which coincides with the successful observation of compressibility) corresponds to a reduction in the volume of the object of more than 1%.*

On the other hand, we must also conclude that, following the observation, the entity has in fact acquired the property of incompressibility. Indeed, if we would decide to carry out again the observational test, we can predict with certainty that the outcome would be positive this time! This is because to actually observe the incompressibility of the object we had to compress it, and a non-elastic object, when compressed, automatically becomes incompressible (according to our definition of incompressibility). In other words, we must conclude that the observational process has created the same property it was meant to observe! (See Figure 21).

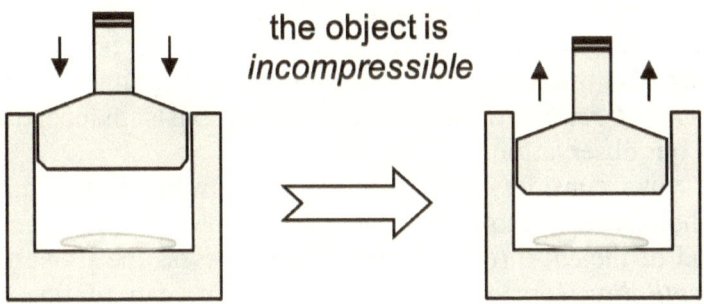

Figure 21. *An object that was previously compressed becomes incompressible.*

This may seem a little bizarre: the failure to observe the incompressibility has, as its effect, the creation of incompressibility! Of course, it is possible to imagine examples in which the observation needs not to be unsuccessful in order to create what is observed. But here we must be clear about what we exactly mean by the word "create."

In fact: *one thing is to create a property and another one is to merely highlight its existence!* Sometimes the process of highlighting the existence of a property can be interpreted as a creative process, although strictly speaking it should not be considered as such. This confusion arises because we tend to consider the properties of the physical entities in *static*, rather than in *dynamic*, terms.[33]

To better understand what we mean by this, consider again the example of the match, which, evidently, has the property of being ignitable even before the observer chooses to observe it, by striking it against a suitable surface. *But for what reason can we affirm that an unburned match possesses such a property?*

Simply because we can *predict with absolute certainty* that if we would strike the match, this will fire up. It is based on this certainty in our prediction of the outcome of the observational

---

[33] Baltag, A. and Smets, S., *Quantum logic as a dynamic logic*. Synthese 179: pp. 285–306 (2011).

experiment that we can decree that the match *possesses in actual terms* the ignitability, even when it has not yet been lighted. To use Einstein's terminology, the ignitability is an *element of reality* of the match, which exists independently from our observation.

Let us consider an example borrowed from human psychology. It is known that there are people who, for a certain period of their life (or their entire life) possess the property of *susceptibility* (or *oversensitiveness*). We can define this property as an excessive emotional reactivity of the person when s/he receives a critical judgment. Of course, a susceptible person will not manifest its susceptibility in all circumstances, but only when confronted with a judgment.

One might then be tempted to say that the susceptibility is created only on the occasion of a judgment, and that outside of these circumstances the property would not be attributable to the person. But this would be a logical error. *We need in fact to distinguish a property from the outcome of an observational experiment that may confirm its actuality.* The susceptibility of a susceptible person is a property that this type of person always possesses, in a stable manner, regardless of whether it is or not observed in practice, by means of an adequate observational test.

The crucial point in all this is the *determinability* of the behavior of the person. As already noted by Einstein and his two collaborators, *Boris Podolsky* and *Nathan Rosen*, in their famous article of *1935*,[34] and later on by Piron, Aerts and others:

> *The attribution of a property to a physical entity (in the sense that the property is actual for that entity) is equivalent to the possibility of predicting with certainty (at least in principle) the positive outcome of the corresponding observational test.*

So, if the susceptible person is such, it is because if we would

---

[34] Einstein, A., Podolsky, B., e Rosen, N. "Can Quantum-Mechanical Description of Physical Reality Be Considered Complete?" Phys. Rev., 47, pp. 777-780 (1935).

decide to make a critical judgment against her/him, we can predict with certainty its emotional overreaction. Thus, when we observe the susceptibility, we are not really creating it, but just confirming it.

Of course, the certainty of the prediction is essential for the assignment of a property. In general terms, consider an entity S and an interaction I between S and the instrument M that performs the observation. Suppose that every time the interaction between S and M is switched on, the outcome O inevitably occurs. This simply means that S possesses the property A of producing O whenever it interacts with M according to I. And of course, S possesses such property A even when it doesn't interact with M.

The aim of this long digression is to emphasize that in ultimate analysis the properties possessed by a physical system are statements of a *dynamical* kind, in the sense that they correspond to the way a system reacts when solicited to interact with other systems, according to given modalities. These interactions, subtended by the observational processes, depending on their characteristics may also fundamentally alter the state of the system, but this does not necessarily mean that the observation would have for this created the observed property.

To further clarify this conceptually subtle aspect, consider again the previously defined left-handedness' property, which we now denote in more precise terms as *left-handedness-1*, to distinguish it from the *left-handedness-2* property, which we are going to define. The observational procedure of the left-handedness-2 is very simple and is as follows:[35]

> *Observational protocol of left-handedness-2*: Take the elastic (or the longer fragment of the elastic, in case it would be already broken) and with a scissor cut it exactly

---

[35] For didactical reasons, the present definition of left-handedness-2 differs slightly from that proposed in Sassoli de Bianchi, M. "God may not play dice, but human observers surely do." Foundations of Science 20, pp. 77–105 (2015).

in two pieces, in a way that one of the two fragments is noticeably longer than the other one (see Figure 22). Then, if the band is black, grab the longer fragment with the left hand and the shorter fragment with the right hand; conversely, if the band is white, or of any other color, do the opposite. After that, simply take notice in which hand is the longer fragment: if it is in the left one, then the elastic is, by definition, *left-handed of type 2*, if it is in the right one, it exhibits the inverse property, which is the one of being *right-handed of type 2*.

The difference between left-handedness-1 and left-handedness-2 allows us to clarify in what sense a quantum observational process is able to literally create the observed property. In fact, it is clear that the observation of the left-handedness-2, although it also produces the breaking of the elastic and the creation of a specific relation between its fragments and the hands of the experimenter, not for this can it be regarded as a process through which the observed property is really created. Indeed, it is definitely possible to predict in advance and with certainty, without disturbing the elastic-entity, what will be the outcome of the experiment, since by definition all black elastics possess the property of left-handedness-2, while all white elastics, or of any other color, that of right-handedness-2.

But we cannot say the same for the left-handedness and right-handedness of type 1. In this case, we have no way of predicting the outcome of the observational experiment, and that is why, in ultimate analysis, we can affirm that such observation literally creates the observed property, that is to say that the left-handedness-1 (or right-handedness-1) is not possessed in an actual sense by the elastic prior to its observation, but only in a *genuinely potential* sense. And this is why we can assert that there is a fundamental connection between the concept of *creation* and the one of *unpredictability*.

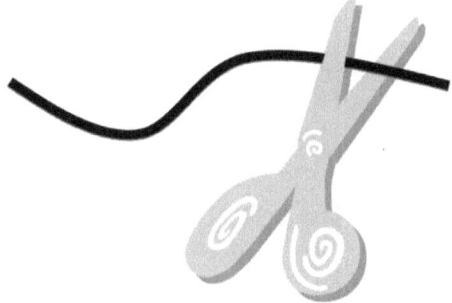

Figure 22. *In the process of observation of the left-handedness-2, the breaking (i.e., the cutting) point of the elastic, and subsequent attribution of the two obtained fragments to the hands of the observer (depending on the color of the elastic) are entirely under the directive control of the latter.*

10. THE TRUE MYSTERY: NON-SPATIALITY

We have reached the last part of our reflection. What we have shown is that the ability to predict with certainty the outcome of an observation (and hence to attribute, in a stable way, the corresponding property to the observed entity) depends on the very definition (in the operational sense of the term) of the observational process, in relation to the observed entity.

In the case of left-handedness-2, the observational process is by definition entirely under the control of the experimenter, and therefore its outcome is predictable with certainty at any time (provided we know the state of the system, i.e., the color of the elastic). Instead, in the case of the left-handedness-1, the way the observational process is designed precludes a priori this possibility of control, and therefore results in an irreducible unpredictability of its outcome.

In other words, to use the typical jargon of quantum physics, the state of a black colored elastic (respectively, of a non-black colored elastic) can be described either as an *eigenstate* of the observational process of left-handedness-2 (respectively, of the right-handedness-2), or as a *superposition state* in relation to the observational processes of left-handedness-1 and right-handedness-1 (regardless of the color).

In fact, since an elastic always actually possesses its specific color, it also actually possesses the left-handedness-2 (if it is black), or the right-handedness-2 (if it is not black); therefore, it is always in an eigenstate (of non-superposition) with respect to these two mutually exclusive properties. On the other hand, since an elastic cannot actually possess (except in specific circumstances, of an ephemeral nature) the left-handedness-1 or the right-handedness-1, in relation to these two properties (which are also mutually exclusive) the elastic is typically in a superposition state, in the sense that it "possesses" them both, conjunctly, *but only in a potential sense*, meaning that both are always potentially actualizable, although not simultaneously actualizable.

The crucial difference between these two situations resides, as already stated, in the possibility or impossibility to operate a full *control* over every aspect of the interaction between the observed system and the observer system. At this point, considering the deep analogy offered to us by the "physics of the elastics," we can ask ourselves the following question, for example in relation to the spatial localization of an electron (or of any other microscopic entity):

> *Considering that the quantum probabilities cannot be attributed to a lack of knowledge of the observer about the specific state in which the electron is, prior to the observation, and that it is not possible to predict in advance, with certainty, not even in principle, the specific spatial localization of the same, what can we conclude about the spatiality of such microscopic entity?*

According to the logic of our analysis, we are forced to conclude that:

> *The spatial localization of an electron is not a property which pre-exists its observation; therefore, contrary to the ordinary objects of the macroscopic world, an electron is not an entity usually present in our three-dimensional space.*

In other words, an electron does not possess in actual terms a specific localization, except in the moment it is detected by an instrument of observation, in the same way as an elastic band does not actually possess a specific *lateralization* (of type 1), except when such a lateralization is actualized (created), although ephemerally, through an observation. This is simply because, as we already emphasized, there is no way to predict with certainty its localization, and therefore the same criterion of existence of such a property no longer applies.[36]

---

[36] This is the case, for example, because of *Heisenberg's uncertainty principle*, which prevents us to observe simultaneously the position of

It is important at this point to draw a distinction between the concept of "being *present* in a region of space" and the concept of "being *detectable* in a region of space." In fact, the non-spatiality of the microscopic entities, i.e., their *non-locality*, does not imply the impossibility of their spatial detection (see Figure 23).

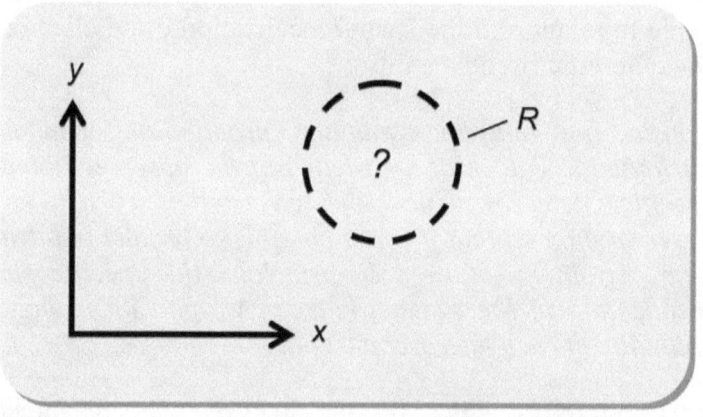

Figure 23. *Although a microscopic entity is generally always detectable in any region R of space, with a certain probability, it is not possible to conclude from this that prior to the detection the entity was present in that region.*

the electron and the way such position varies locally in time, i.e., its velocity. And if we cannot jointly determine the position and the velocity of the microscopic entity, neither can we solve the classical equations of motion, which would allow us to determine the spatial trajectory that the entity goes along in time, and therefore its successive positions. If this is not possible, the concept itself of a spatial trajectory associated to the microscopic entity has to be abandoned. There are of course many other reasons we could evocate to conclude about the evident *non-spatiality* of the microscopic entities, but we cannot recall them here, not to overly extend this booklet.

The currently known microscopic entities, such as the electrons, protons, neutrons, atoms, etc., while being non-spatial, are nonetheless *fully available in relating to the macroscopic entities that form our ordinary three-dimensional space*, in the sense that they are always available in manifesting their presence – however ephemeral – in that space, in the ambit of an observational process, i.e., in the ambit of their interaction with a macroscopic entity that acts as a measuring instrument.

This means that although there is no way to predict with certainty whether the observation of an electron (of which we know completely the state) in a given region of space will give a positive outcome or not, we can nevertheless predict with certainty that if this region would be expanded to cover the entire three-dimensional space, then the probability of detecting its presence would be equal to *100%*.[37]

We can express this idea in a somewhat more precise way by partitioning the three-dimensional physical space into two arbitrary regions: R_L (left region) and R_R (right region), separated for example by a two-dimensional infinite plane (we are proposing here a *gedankenexperimente*, that is, a so-called *thought experiment*; see Figure 24).

We can then define the property of *localization in R_L* (respectively, of *localization in R_R*) of an electron as the property of the same to be detected without fail in R_L (respectively, in R_R).

Here we find ourselves in a situation very similar to that of the experiment of left-handedness (of type 1) of the elastic. Indeed, as well as the elastic band is *100%* available to interact with both hands of the human observer, in order to actualize either the left-handedness-1 or the right-handedness-1 property, in the same way an electron is *100%* available to interact with both spatial regions R_L and R_R (more precisely, with the position detectors which are placed in them), so as to express either the property of localization in R_L, or the property of localization in R_R.

---

[37] In the mathematical formalism of the theory, this fact is expressed by the condition of *normalization* of the wave function describing the state of the electron.

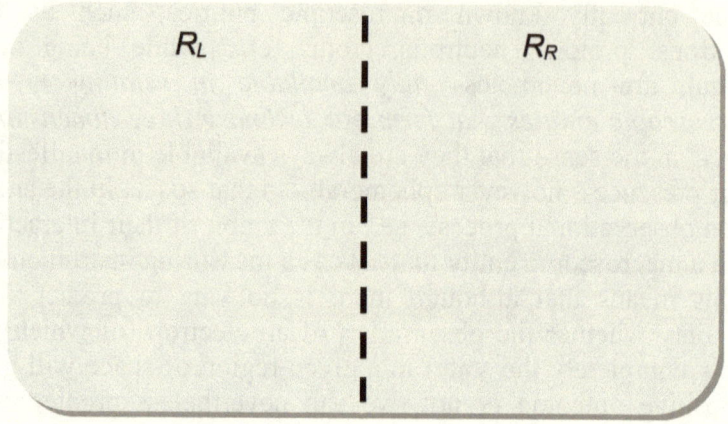

Figure 24. *A (two dimensional) symbolic representation of the (three-dimensional) physical space, partitioned into two distinct regions (dotted line): a left region, R_L, and a right region, R_R.*

However, the localization is literally *created* at the very moment of its observation, and therefore it doesn't make sense to say that before the observation the electron was factually (i.e., actually) already present in one of the two regions. The only thing we can say is that it was *potentially* present in both regions, as well as an elastic band is potentially both left-handed and right-handed (of type 1).

But we certainly cannot say that the electron, before the observation, was *actually* in both regions, as this means that it would be detectable with certainty, simultaneously, in both regions, whereas an electron can only manifest, when observed, in a single spatial location at a time (as well as a left-handedness' experiment can only yield a positive or negative outcome, but not both outcomes simultaneously).

Now, as we must not confuse the full availability of an elastic band (for the very fact of its existence) to participate in a left-handedness-1 experiment, with the possible outcomes of the experiment, so we must not confuse the full availability of an electron (for the very fact of its existence) to participate in an experiment of spatial localization, with the possible outcomes of the experiment. Because while it is true that an electron is always observable in our physical three-dimensional (ordinary) space, that is, detectable in it with probability equal to *100%*, this doesn't mean that it possesses a specific localization inside of it, i.e., a specific position.

We have here a typical example of a situation in which the *hypothesis of reductionism* does not apply anymore. While on one hand we can affirm that the electron is *globally present* (in the sense of being detectable with certainty) in the entire three-dimensional physical space, at all times, on the other hand we cannot conclude that for this it is *locally present* in one of its regions, i.e., in one of its parts. And this means that in general the nature of the relationship of an electron with the totality of the three-dimensional space cannot be deduced from its relationship with its parts. Not in general if nothing else.

And of course, here we approach the true mystery of the microscopic world, which turns out to be populated by entities whose *non-spatial* nature is quite different from that of macroscopic entities, i.e., from the world of everyday objects, and therefore cannot be described and understood in the same way.

When we are dealing with microscopic entities, such as electrons, the metaphor that consists in thinking of our three-dimensional physical space as a *container* no longer applies. Or, rather, it does not apply if we consider the observation of the spatial position of an electron in the usual sense of a process through which the electron would reveal an already acquired presence in a given region of space, and not as a process through which an electron would be in a sense *forced* to manifest in that region.

By this, we mean that because an electron is fully available in *globally* manifesting itself in spatial terms, it is always possible to exploit such a full global availability to confer to its manifestation a specific, predetermined outcome. We can do this in the same way as we can exploit the full availability of an elastic band in letting itself be cut (in any point), to define a different concept of left-handedness (or right-handedness), that we have called left-handedness-2, such that we can predict the observational outcome in advance, and thus corresponds to a property actually possessed (or actually non-possessed, depending on its color) by the elastic band, even before its observation.

For example, in principle we can always apply in region R_R a *highly repulsive force field*, so that the probability of the electron to be detected in this region will tend to zero, as the intensity of the repulsive field increases. In other words, by increasing the repulsion of the applied field in region R_R, it is possible to proportionally increase the *control* of the observer over the spatial detection experiment of the electron in region R_L (see Figure 25).

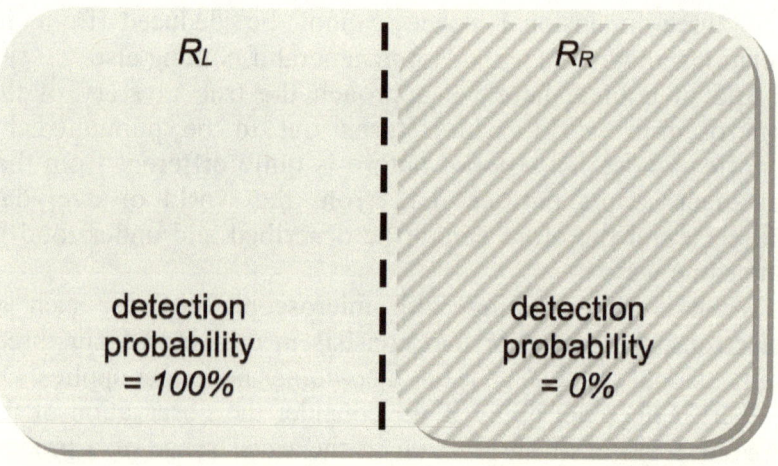

Figure 25. *The application of a highly repulsive force field in region R_R prevents the electron to be detected inside of it.*

In other words, along the lines of the definition of left-handedness-2, we can define the property of *localization-2* of an electron (and of any other physical entity) in the following way:

> *Definition of localization-2*: A physical entity is said to possess a *(spatial) localization of type 2*, in a given spatial region R, if it can be detected without fail (that is, with

probability equal to *100%*) in that region, when in the complement region a virtually infinite repulsive field is applied.[38]

According to this definition, it is perfectly licit to affirm that an electron always possesses in actual terms the *location-2* property, for instance in the region R_L, as is clear that we can always predict with certainty that, if we would carry the observational experiment, the electron would inevitably be detected (i.e., localized) in R_L.

Of course, by varying the region of application of the repulsive force field, for example moving it from region R_R to region R_L, the electron would automatically lose its localization-2 in R_L and acquire it instead in R_R; in that sense, the property of localization-2 should be more properly called *localizability-2*.

We can therefore assert that an electron certainly always actually possesses a *type 2 localization* (or *localizability*), whereas in general it doesn't possess a localization in the ordinary sense, which we shall denote *localization-1,* to distinguish it from *localization-2*.[39]

The observation of localization-2 is a process that, like the observation left-handedness-2, is entirely under the control of the experimenter, and therefore the outcome of the observation can be predicted with absolute certainty. This means that localization-2 is a property that a microscopic entity such as an electron possesses in a stable way, independently of whether it is or is not observed, i.e., made manifest.

It expresses the full availability of such entity (and of any other known microscopic entity) to relate with the structure of our three-dimensional space *as a whole* (and more exactly with the

---

[38] To avoid unnecessary complications due to the mechanisms of creation of antimatter from a high energy field (*Klein paradox*), we limit our discussion to a purely non-relativistic ambit.

[39] A physical entity possesses an (ordinary) localization, of type 1, if it is present with certainty in a given region of space, regardless of our observation, in the sense that it is always possible, at any moment, to identify a sufficiently large spatial region, such that the entity would be detectable with certainty in its interior.

macroscopic entities that characterize it), thus also its complete availability to remain "confined" in a specific region of the same (in the sense of being detectable with certainty in that region), when its manifestation in any other region is being prevented.

In other words, we can affirm that localization-2 is that weaker form of spatial localizability possessed by the microscopic entities, which on the basis of their *availability to relate* with the three-dimensional space as a whole allows them to relate exclusively with a specific part of this space, when by means of some kind of active control the possibility for them to relate with the other parts of space is prohibited.

Now, since to relate to the three-dimensional space means, in ultimate analysis, to relate with the entities that populate it stably, that is to say with the macroscopic entities that characterize it, localization-2 is nothing but that property allowing a microscopic entity to *stably bind* to a macroscopic entity (see Figure 26), that is to form what is usually referred to as a *bound state*, so acquiring the same type 1 spatial localization of the latter.

In other words, when a microscopic entity binds to a macroscopic entity, that is, it becomes part of it, not only it makes manifest the property of localization-2, but in fact also acquires (i.e., actualizes) the localization-1 property, possessed by the macroscopic entity. Exactly in the same way as an elastic band, when it manifests the left-handedness-2, obviously also manifests, in the same moment, left-handedness-1, since the successful outcome of the experiment characterizing left-handedness-2 is the same as the one determining left-handedness-1.

Thus, we can affirm that the microscopic entities are entities that usually do not possess the property of belonging to the ordinary three-dimensional (*Euclidean*) space, as they do not actually possess, permanently, the localization-1 property, which is possessed instead by the macroscopic entities. On the other hand, because they do possess the localization-2 property, which is an expression of their full availability in binding with specific macroscopic entities, they can nevertheless enter and reside permanently in the ordinary physical space, in the form of *aggregates* (see Figure 26).

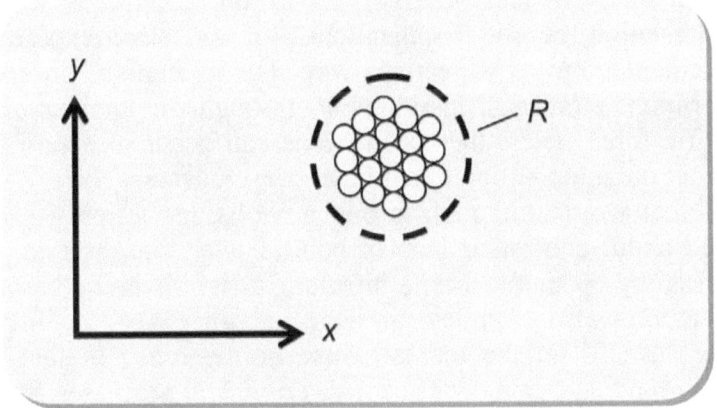

Figure 26. *A macroscopic body is an aggregate of microscopic entities (improperly represented here as corpuscles), which possesses the property of localization-1 (in this case in the region R). The same is true, in a sense, for its components, at least for as long as they remain in their condition of intimate interconnection.*

But if a macroscopic aggregate stably possesses the property of spatiality, we must be careful not to think of its individual components as entities that would also possess the same property. Contrary to the hypothesis of reductionism, it is not the spatiality of the microscopic constituents that confers to the macroscopic body its spatiality, but the other way around.

The spatiality of the macroscopic body is in fact an *emergent* property, that the individual components only possess when all together, for as long as they maintain their specific and exclusive relationship, but they immediately lose as soon as they separate, or get separated, from the aggregate in which they belong.[40]

---

[40] This fact, as shocking as it may be to our spatial and corpuscular intuition, which was formed from macroscopic objects, is nevertheless quite natural if we think in more abstract terms: an aggregate, in fact, precisely because it is such, necessarily possesses properties which are different from the properties of its individual components.

Of course, we can observe that as the conditions for the manifestation of the localization-2 of an electron can be implemented in a directive way by a human observer-experimenter (at least in principle) through the application of specific force fields, these conditions can occur in nature also without the intervention of a human consciousness.

In other words, although in our previous discussion we have talked about control or lack of control over the observational process by a human experimenter, it is obvious that this distinction also applies in the absence of a human consciousness, in the mental sense of the term. In fact, we simply have to distinguish the *indeterministic* processes, with a built-in "symmetry breaking" mechanism sensitive to the tiniest fluctuations in the environment, from the *deterministic* ones, which on the contrary are not sensitive to these fluctuations.

After all, the human observer simply adds to the environmental physical processes, either as a *passive discoverer* of them, or as an *active participator*, when s/he *creates* specific *interrogative contexts* in order to implement precise operational questions, and acts accordingly upon the entities subject of her/his investigation. But of course, these experimental contexts will manifest their results regardless of whether there is ultimately a human mind to take cognizance of them, as in fact it almost always happens in modern physics laboratories, in the ambit of procedures which are often fully automated.

To give a very elementary example, a certain quantity of liquid formed by the aggregation of *1000* drops of water obviously has a volume V *1000* times greater than the volume v of a single drop. In other words, the liquid, as an aggregate, is characterized by the property of possessing a volume $V = 1000 \times v$, and all the water droplets that form it also possess, in a sense, this same property, as it is not possible to say where in the aggregate a drop ends and the other begins. On the other hand, if taken separately, that is, if separated from the quantity of liquid, the individual drops no longer participate in such property.

11. CONCEPTUAL ENTITIES

We arrived at the penultimate chapter of our short text (which in its first edition was the last one), in which we have attempted to clarify the measurement problem of quantum physics, namely the nature and origin of that observational process (called a *process-1* by von Neumann) that allows to switch from a description in terms of probabilities (not due to a lack of knowledge about the system), to the concrete realization of a specific event, that is of a phenomenon factually observed in the laboratory.

For the purpose of this clarification, we have illustrated, by means of an extremely simple example, the fundamental ideas of the so-called *hidden-measurement approach*, conceived in the last decades by Diederik Aerts; an approach which in turn is part of a wider explanatory framework, named the *creation-discovery view*, from which emerges the possibility of a more mature and articulated form of realism in comparison to the naive view of classical realism, which can take into account the different modalities through which an observer is able to actively (and often invasively) interrogate the physical systems, attributing to them not only actual properties, but also potential properties (i.e., properties that are available to be actualized, although not in a predeterminable way).

Thanks to this analysis, it becomes quite evident, at least for this writer, that the *quantum observer effect* has no reason to be understood as a psychophysical effect, where abstract possibilities would be reduced into concrete actualities by a human mind, being rather an expression of a purely physical mechanism of symmetry breaking, always at work when the observational process does not contemplate the possibility of a full control of the interaction between the *physical* observed-system and the as *physical* observer-system.

We also sought to clarify that the true great mystery of

quantum physics is not so much the observer effect, but the understanding of the genuinely *non-spatial* nature of the entities of the microworld, which, although they are able to stably bind to the macroscopic entities, not for this they can singly and stably belong to the ordinary spatial theater of the human experience. There exist in fact a *pre-spatial* – and thus also *pre-temporal* – level (or layer) of the physical reality, populated by microscopic entities (and probably also by other entities, as yet unknown) from which emerged the macroscopic bodies, in a way which still needs to be clarified.

A major difficulty in the understanding of the emergence of macroscopic entities (and of the three-dimensional space that contains them) from the pre-spatiotemporal level of microscopic entities, could be our insistence in thinking about the latter in mere *objectual* terms. Still, one may wonder: what other models would we have at our disposal to mentally visualize the entities of the microworld, which in addition to their staggering lack of spatiality present a number of other oddities, like their well-known lack of *distinguishability*, their tendency to produce continuous *interferences*, form connections of any kind regardless of spatial distances (*entanglement*, see the next chapter), and so on?

A possible answer comes to us once again from the pioneering work of Aerts and collaborators. Indeed, within the broader explanatory framework of the *creation-discovery view* is contained not only the possibility of a refined conceptual analysis, but also a powerful mathematical model, capable of describing, as already mentioned, both classical (phase spaces) and quantum (Hilbert spaces) structures. More recently, this already quite extensive mathematical and conceptual framework has been further expanded in order to embrace an even more general class of systems, called *SCoP* (*state-context-property systems*), thanks to which it becomes possible to describe not only the action of an experimental context on a given system (as is usual in physics), but also the influence of the system on the context itself.[41]

---

[41] Aerts, D. "Being and change: foundations of a realistic operational formalism." In: D. Aerts, M. Czachor and T. Durt (Eds.), *Probing the*

One advantage of such a general approach is to allow the description not only of physical entities (classical, quantum or quantum-like), but also of more abstract entities, such as *human concepts*, *human minds*, and the *decision processes* associated with them. This has allowed to discover that the generalization of the formal structure of quantum mechanics lent itself surprisingly well to the construction of a quali-quantitative theory of human concepts and their combinations.[42]

Obviously, we cannot enter here into the details of this theory, which is very articulate and would require an additional booklet just to introduce its basic concepts. But what we want to highlight here is that following the discovery that the quantum formalism was perfectly able to model human concepts, a rather unusual, but not less natural, idea quickly emerged, summarizable in the following question:[43]

> *If quantum mechanics, as a formalism, models human concepts so well, perhaps this indicates that quantum entities themselves are conceptual entities?*

This fascinating question has given rise to a very innovative interpretation of quantum physics, which is based precisely on the assumption that the *nature* of quantum entities would be *conceptual*, in the sense that these entities would interact with the macroscopic measuring instruments (and more generally with the entities made of ordinary matter) in a similar way as the human concepts interact with human minds (or other

---

Structure of Quantum Mechanics: Nonlinearity, Nonlocality, Probability and Axiomatics. Singapore: World Scientific, pp. 71–110 (2002).

[42] Aerts, D. and Gabora, L. "A theory of concepts and their combinations I: The structure of the sets of contexts and properties." Kybernetes, 34, pp. 167-191 (2005); "A theory of concepts and their combinations II: A Hilbert space representation." Kybernetes, 34, pp. 192–221 (2005).

[43] Aerts, D. "Quantum particles as conceptual entities: A possible explanatory framework for quantum theory." Foundations of Science, 14, pp. 361–411 (2009).

memory structures sensitive to the meaning of the concepts).

As is known, human concepts are typically non-spatial entities. In fact, one can hardly say that they belong to our three-dimensional space, but rather to a *mental space, abstract* in nature. Of course, the mental space of human concepts, in the classical view of materialism and reductionism, originates precisely in the activity of human brains, which apparently are contained in the ordinary spatial theater. On the other hand, independently of the fact that human concepts originate or not from our specific localized brain-structures, the fact remains that the spatiality of a human concept, such as for example the concept "fruit," is quite different from the spatiality of an ordinary physical object.

In fact, the concept *fruit*, being an abstract entity, it doesn't actually possess, in a stable way, the property of localization-1, which is instead typical of concrete objects, while it unquestionably possesses the property of localization-2, since the fruit-concept is always fully available in interacting and binding with *specific semantic entities*, for instance formed by *specific aggregates of concepts*, in the context of specific phrases.

In other words, in the same way an electron is able to relate/bind (in an ephemeral or stable way, depending on the type of interaction) to a specific macroscopic system, thus manifesting its presence in the three-dimensional theater, also the human concept *fruit* can temporarily assume the status of an *object*, when it relates/binds with a specific *objectifying* context, for instance in the ambit of the following injunctive sentence:

| Look at the *fruit* that is right now on the table!

If on the table in question a single apple is for instance present (see Figure 27), we can consider that the observation of the localization of the conceptual entity *fruit*, in the (experimental) context of the above sentence, is entirely predetermined.

Figure 27. *The "fruit" abstract concept is stably objectified in the concrete concept of the single apple present on the table, thanks to the experimental context expressed by the injunctive phrase "Look at the fruit that is right now on the table!"*

This is because the presence of a single apple on the table *forces* the concept *fruit* to relate/bind, via the *eye-brain* instrument of the human observer, exclusively to that particular apple. We are therefore in the ambit of what we have called *localization-1*.

On the other hand, if on the table there were *two apples* (see Figure 28), the experimental context would no longer be able to force the concept *fruit* to assume a predefined spatial localization, so that the human entity that has to act the above-mentioned injunction must *choose* which of the two apples to look at (i.e., select a specific visual interaction), attributing in this way to the concept "fruit" an *ephemeral* spatial localization, which will correspond to the localization of the apple chosen in that moment (and which may change in a subsequent moment). And since the decision process is usually quite sensitive to the *intrapsychic* and *extrapsychic* fluctuations, in no way it can be predicted in advance. Therefore, we are now in the ambit of what we have called *localization-2*, a property that the abstract concept *fruit* doesn't ordinarily possess in actual terms, that is, permanently.

Figure 28. *The "fruit" abstract concept is ephemerally objectified into the concrete concept of the apple-1 on the left, or the apple-2 on the right, in a way which is generally unpredictable.*

The non-spatiality of human concepts is, therefore, quite similar to the non-spatiality of quantum microscopic entities. However, it's important to emphasize that the analogy between human concepts and microscopic entities, as demonstrated by the careful analysis of Aerts (in what is today called the *conceptuality interpretation* of quantum mechanics), is much deeper and articulated than what could be understood by the abovementioned example. In fact, human concepts are able to express almost all the complex phenomena typical of the microscopic quantum level, as *entanglement*, by means of the violation of *Bell's inequalities*, *superposition* and *interferences*, *coherence* (which in the ambit of human concepts becomes *connection through meaning*; see the next chapter), *incompatibility* (expression of the impossibility for a concept of being simultaneously maximally abstract and concrete), etc.[44]

On the other hand, while taking into account these similarities, it is important not to fall victims of too easy *anthropomorphisms* and confuse human concepts with the microscopic (or even macroscopic) quantum entities, or human minds with the measuring apparatuses in a laboratory. The same example that we have just given, of the concept *fruit* that is objectified in a *physical apple placed on the table*, could in this sense be slightly misleading. In fact, it might suggest to the reader that there is no difference between the concepts elaborated by a human mind and the physical entities, but obviously this is not the case.

Think of an *acoustic wave* and an *electromagnetic wave*. Obviously, these are completely different physical entities. On the other hand, they share the same *wavy nature*, and this means that under certain circumstances they will be able to have similar behaviors, for example in the context of the so-called *interference phenomena*. The same is true of physical entities such as an electron, which would share with human concepts the same *conceptual nature*, while remaining entities that are perfectly distinct from the latter.

Historically speaking, we humans have "constructed" our conceptual world by *abstracting* it from the objects of our everyday life. This means that the so-called objects have been associated with the most concrete examples of our human conceptual reality. But this is only due to the way we interacted with the objects of our environment during our evolution on the surface of our beautiful planet. In fact, this has played an important role in shaping our language, that is, in creating more abstract concepts, for example when we felt the need to indicate an entire category of objects, instead of just one object, i.e., a

---

[44] Aerts, D., Sassoli de Bianchi, M., Sozzo, S. & Veloz, M. (2018). On the Conceptuality interpretation of Quantum and Relativity Theories. *Foundations of Science*. https://doi.org/10.1007/s10699-018-9557-z.

specific member of a category of objects.

It follows that we can identify a *human historical line* that allows us to go from the concrete to the abstract, where the most concrete concepts are the entities of a spatio-temporal nature, the objects precisely. When we say "this apple that I am holding right now," we are using a maximally concrete concept, whereas when we say "entity," "thing," or "element," we are using concepts that are maximally abstract, and of course between these two extremes we can insert concepts of an intermediate level of abstraction or concreteness.

This human "parochial" line of ours, which allows us to go from the concrete to the abstract, is the one that is taken into account today in psychology, or in the field of semiotics. But there is also a second line, which we can consider to be of a more universal nature, which also allows us to move from the abstract to the concrete, and is linked to the possibility of combining together different concepts, to create more complex emerging meanings.

According to this second line, the most abstract concepts are simply those expressed by single words, while the more concrete are those that we describe using a large number of interconnected terms, which in our human language correspond to what we commonly refer to as "stories." A story is an aggregate of concepts combined with each other according to a specific narrative, that is, according to a specific meaning, and the macroscopic objects, if we consider them to be formed by (non-human) conceptual entities, would be the equivalent in the material world of what we call stories in the human conceptual world.

And indeed, macroscopic objects and stories have similar behaviors. Take an object A (such as an apple) and an object B (such as a pear). If you consider the conceptual combination "*A and B*", using the logical connective "e," you are still able to put it in correspondence with an object, and more precisely with the object obtained by putting the two objects together (the apple and the pear), which thus form a single *composite* object.

In other words, if A and B are two objects, "*A and B*" is still an object. On the other hand, when we consider the conceptual combination "*A or B*," using the logical connective "o," we are no longer able to associate it with a specific object. In other words, if A and B are two objects, "*A or B*" is no longer an object, but only a concept.

The conceptual world, unlike the objectual world, is therefore "closed" with respect to the operations of the logical connectives of conjunction and disjunction, while the world of objects is only closed relative to the conjunction operation. The same goes for the stories, that is, for the conceptual entities formed by very large combinations of concepts that are connected to each other. If A and B are two stories, then "*A and B*" can be considered in turn to be a story: the story that we can tell by first reading A and after B.

But similarly to spatiotemporal objects, if A and B are two stories, "*A or B*" will not usually be considered to be a story. We can see this for example in the fact that we can easily find collections of different stories, within a same book, but we will hardly find stories that are the conjunction of two long narratives (although in certain particular contexts this is possible, for example when a detective takes into consideration various hypotheses about how the facts could have taken place).

Naturally, it is not possible to go here into the details of the many subtleties of the *conceptuality interpretation*. What we wanted to stress is that there is the possibility to understand both the microscopic quantum entities and the classical material objects as entities of a conceptual nature, the latter being a vast combination of the former, whose emergent behavior is such that it becomes extremely difficult to put them into a state of superposition, expression of a sort of "logical conjunction," in the same way that it is difficult to find narratives formed by the superposition of other narratives, when they become too complex and articulated.

It is crucial, however, to always keep in mind that a human concept, even when it indicates a concrete object, and therefore it is a concept that is maximally concrete, it must not to be confused with the object with which it is put in correspondence, although also the object in question, from the viewpoint of the conceptuality interpretation, would be an entity of a conceptual nature, in a state of maximum concreteness (the equivalent of a long story).

What the profound analysis of Aerts teaches us is simply that human concepts would not be the only *conceptual entities* with which we humans have to do, since also the entities of the microscopic world, like electrons, and their aggregates, would have that typical behavior that we humans usually attach to concepts, and not objects.

We can therefore observe that the approach to quantum physics which was born in Geneva and matured in Brussels, at the *Center Leo Apostel*, while on one hand it was able to demystify the observer effect, showing that the human mind plays no specific role in the *actualization of potential* of microscopic systems (apart of course the fact of conceiving and creating particular experimental contexts), on the other hand it showed, thanks to the generality of its approach, the existence of *mental-like* dynamics of a *non-human* origin. In the sense that the interactions of the microscopic entities with the macroscopic ones are no doubt better described by terms such as "communication, language, concepts, symbols, meanings, minds, memories, etc.," instead of terms like "objects, bumps, bounces, collisions, forces, waves, etc."

In other words, in its long journey of scientific investigation of the physical reality, human beings, perhaps for the first time, must surrender to the evidence that although a reality exists "out there," independent of our mental representations of it (unless evidence to the contrary), such reality is structurally much more similar to the very minds that explore it, than what we could initially suspect. In the sense that:

If we truly want to understand the nature and behavior of the matter-energy, at a fundamental level, like it or not we cannot avoid thinking of its attributes in mentalistic terms: to understand it we need to psychologize it, although not in the sense of a mere human psychology, and certainly without falling into too easy mysticisms and anthropomorphisms.

12. CONNECTIONS AND DISCONNECTIONS

In this last chapter, which we considered adding to the new edition of the book (in addition to the next short chapter on the "other observer effects"), we would like to describe, and in this case partly demystify, another important phenomenon related to the *non-spatiality* of the entities of the microworld: the so-called *quantum entanglement* (which the French indicate with the term of *quantum intricacy*).

The "entanglement problem" appeared very early in the course of the development of quantum theory, although initially only as a problem of exquisitely theoretical nature, linked to the correct interpretation of the formalism and its alleged inability to fully describe any possible experimental situation.

We have already mentioned in Chapter 9 the reality criterion initially enunciated by Einstein, Podolsky and Rosen (in short, EPR), in their famous article of 1935, in which they were able to obtain a contradiction by reasoning on a particular quantum system. In Chapter 4, we also referred to a result by *Diederik Aerts*, who endorsed Einstein's initial suspicion that quantum theory was an incomplete theory.

Einstein's and collaborators' reasoning focused on a particular class of systems, called *bipartite systems* (that is, formed by two parts), when they are in a particular state, called an *entanglement state*. In this chapter, we will try to explain how EPR managed to highlight a contradiction in the quantum formalism, and what the meaning of this contradiction is. Along the way, we also hope to be able to explain the true nature of quantum entanglement, and for what reason it is not possible to understand it without resorting to the notion of non-spatiality.

Consider two bodies, which we will indicate with the letters A and B, and suppose that they move in space, moving away from each other. Suppose also that two experimenters, let us call them Alice and Bob, measure at the same time the respective positions and velocities of these two bodies. Since at the moment of execution of their measurements Alice and Bob are separated by a certain spatial distance, which can be arbitrarily large, they will generally not expect to observe correlations between the results of their respective measurements. However, this possibility cannot be ruled out: it all depends on the history of the two bodies in question.

In fact, if the two bodies were connected to each other in the past, the physical process that caused their disconnection may have created some correlations, which later Alice and Bob can highlight.

A paradigmatic example is that of a rock initially at rest, say located at the origin of the coordinate system of a laboratory, which at some point explodes into two separate fragments, let us call them *A* and *B*, which to simplify the discussion we will assume they have identical masses (see Figure 29).

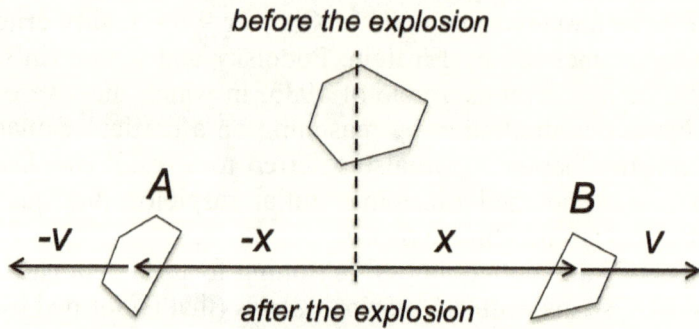

Figure 29. *A rock initially at rest explodes in two fragments A and B of same mass, which then move away in space with opposite velocities.*

The positions and velocities of these two flying rock fragments will necessarily be perfectly correlated, due to the *conservation of momentum*: if in a given instant the position and velocity of the (center of mass of) fragment *A* are *x* e *v*, respectively, then the position and velocity of the (center of mass of) fragment B, at the same instant, will be $-x$ and $-v$.

This situation of perfect correlation is clearly the consequence of how the two fragments have emerged from a single undivided entity, initially at rest, and not the result of some strange connection that would be maintained between them during the course of their journey in space, moving away from one another, or of an imaginary as much as unnecessary communication through which they would agree in every moment how to coordinate their respective positions and velocities.

In their famous article, EPR do not describe the situation of two fragments of a rock that fly away in opposite directions, which would not present any mystery, but that of two entities of a microscopic nature, as could be two electrons, that would interact with each other in the past and that, following their interaction, just as for the two fragments of rock, would find themselves at a considerable spatial distance from each other, in the sense that the probability of observing them in the region where they initially interacted would be essentially zero, while the probability of observing them at a certain distance from that region – distance that increases over time – would be maximal.

Unlike the macroscopic rock fragments, for the two electrons (or photons, or other entities of microscopic nature, the reasoning applies regardless of the type of entities taken into consideration) the quantum laws apply. Now, according to these laws, and more exactly as predicted by the famous *Schrödinger equation* (which we only mentioned in Chapter 2), the two electrons will be in an *entangled state*, and according to this state the results of the measurements made simultaneously and separately by Alice and Bob on the two electrons will necessarily be correlated, as in the case of the two rock fragments that emerge from an explosion.

Why did the presence of such correlations present a problem according to EPR? Let us try to clarify this by explaining exactly what their reasoning was. To begin with, they hypothesized that Bob, by intercepting one of the two electrons (let us call it the electron B), could have measured its position. Now, assuming that following his measurement the result obtained was x, since the two electrons are entangled, it is possible to predict that if Alice had measured the position of her electron A, with certainty she would have obtained a value opposite to that of B, that is, $-x$.[45]

The certainty of such a prediction allows Alice to conclude that at that moment the position of electron A is exactly $-x$, without the need to make any concrete measurement (see our discussion of Chapter 9, where we have highlighted that the attribution of a property is linked to the possibility of predicting with certainty the positive result of a corresponding observational test).

On the other hand, continue EPR, Bob could have decided to measure the velocity of electron B, rather than its position. If he did, he would have obtained a certain value, say v. Also in this case, reasoning as above, Alice could have predicted the velocity of electron A, as also in this case the particular state of entanglement in which the two electrons are in allows her to deduce that if the velocity of B is v , then the velocity of A must be $-v$, just as in the situation of the two fragments of rock.

But here is the final piece of reasoning that EPR have formulated in their famous article. Since the two electrons A and B are separated by an arbitrarily large spatial distance, and since it is quite natural to suppose (it was so at least in the days when they wrote) that such *spatial separation* necessarily implies also an *experimental separation*, the measurements made by Bob on electron B, whether they are position or velocity measurements, in no way they could have influenced the state of electron A. But if this is true, then it is possible to conclude that A possesses not only a well-defined position, at a given instant, but also a well-defined velocity, at the same instant, since Bob is perfectly free to choose which quantity to observe and that his choice is not able to influence the condition

---

[45] Here of course you must trust what we are saying, that is, that these are exactly the predictions of quantum theory, for bipartite systems in an entangled state. It should be pointed out that there are an infinite number of possible entangled states, with very different characteristics, and that the reasoning in question presupposes that the entangled state of the two electrons is of a symmetric kind.

of Alice's electron. This was at least the hypothesis of EPR.

If the above is true, we have an obvious contradiction. In fact, always by using the quantum formalism, it is possible to derive the well-known *Heisenberg indetermination principle*, which prohibits to simultaneously actualize both the position of a quantum entity and its velocity. In other words, the "reality criterion" enunciated by Einstein and collaborators, according to which *the physical properties are nothing but states of prediction* (in the sense that a property is actual if it is possible to predict with certainty the outcome of a corresponding observational test), which is undoubtedly a criterion of very general validity, "in agreement with classical as well as quantum-mechanical ideas of reality," produces a logical contradiction, a sort of paradox from which EPR concluded that quantum mechanics was necessarily incomplete.

The incompleteness in question would result from its alleged inability to describe the simple fact that an electron would always and jointly possess a well-defined position and velocity, hence incomplete as unable to describe every possible property (every possible "element of reality," Einstein would say) associated with a physical entity like an electron.

For many years the uncomfortable question raised by EPR fell substantially into oblivion. There was a swift reaction from *Niels Bohr*, in a rather hermetic article of the same year,[46] with exactly the same title as the EPR article (the title was: "Can quantum-mechanical description of physical reality be considered complete?"), which simply asserted that their whole reasoning was invalid because, according to Bohr, in quantum physics the very concept of "element of reality" did not apply, or something along these lines.

There was another difficulty, however, which partly explains the subsequent lack of interest of physicists with regard to the EPR paradox: the fact that no one believed it was possible to conceive some experiments capable of confirming or refuting the hypothesis of incompleteness posed by Einstein and his young collaborators.

---

[46] N. Bohr, "Can quantum-mechanical description of physical reality be considered complete?" Phys. Rev. 48, pp. 696–702 (1935).

The situation changed radically about thirty years later, thanks to the deep insights of the British physicist *John Bell*,[47] who surprised everyone by deriving some particular mathematical inequalities that today bear his name (*Bell's inequalities*), whose violation would confirm the quantum predictions relative to the phenomenon of entanglement, in experimental situations similar to those described by EPR in their article.

The great ability of Bell (who did not receive the Nobel Prize for his work, but would have certainly deserved it) was to devise mathematical relationships only involving quantities (probabilities) that were directly calculable by using data obtainable in experiments that were in principle feasible.

The other salient feature of Bell's inequalities was that they were able to demarcate the previously described situation of the two rock fragments, from the situation of the two electrons in a quantum state of entanglement. The problem, in fact, is that in both these situations we could observe correlations between the different outcomes of the measurements, but obviously the nature of these correlations was not at all the same.

Diederik Aerts expressed this distinction very well by introducing the following terminology. On the one hand, there are the *correlations of the first kind*, which are those that can only be *discovered* during an experiment. These are correlations that are present in the system even before the experiment is performed, as is the case with the two rock fragments whose velocities and positions remain correlated over time regardless of our observations.

On the other hand, there are the *correlations of the second kind*, which are instead literally *created* by the process of their observation, through the interaction of the bipartite system with the measuring instruments. Only this second kind of correlations are able to violate the famous inequalities discovered by Bell.

About twenty years after Bell derived his inequalities, of

---

[47] Bell, J. (1964). "On the Einstein Podolsky Rosen paradox". Physics 1, pp. 195–200.

which there are many variants today,[48] the first experiments were carried out, by the French group of *Alain Aspect*,[49] which were not only replicated over the years, but performed with an ever greater degree of sophistication, up to the most recent experiments, in 2016, which are believed to have eliminated every possible and imaginable problem of experimental design, which could call into question the validity of the obtained results (these problems are usually called "loopholes," and over the years a considerable number of them have been identified).[50]

So, the experiments confirmed the reality of quantum entanglement, since the obtained data violated Bell's inequalities, which were conceived as we said to act as a watershed between the "classical" correlations, of the first kind, and the "quantum" correlations, of the second kind. Thus, following these experimental successes, most physicists believed that the paradox highlighted by EPR had been solved, in the sense that the experimental results had simply invalidated the EPR reasoning and confirmed the predictions of quantum theory.

But, is it really like that? Not exactly. As we will now try to explain, such a conclusion is merely the result of a misunderstanding about the true, "logical only," nature of the paradox in question. *Diederik Aerts* realized this in the studies he conducted during his doctoral thesis, in the early eighties of the last century.[51]

---

[48] The most famous and widely used is called: *Clauser Horne Shimony and Holt inequality*, or simply *CHSH inequality*.

[49] Aspect, A., Grangier, P. & Roger, G. (1982). "Experimental realization of Einstein-Podolsky-Rosen-Bohm Gedankenexperiment: A new violation of Bell's Inequalities." Physical Review Letters 49, pp. 91–94.

[50] Hensen, B., et al. (2016). "Loophole-free Bell inequality viola-tion using electron spins separated by 1.3 kilometres. Nature, 526, pp. 682–686.

[51] Aerts, D. "The One and the Many: Towards a Unification of the Quantum and Classical Description of One and Many Physical Entities," Doctoral dissertation, Brussels Free University (1981). Vedi anche: Sassoli de Bianchi, M. (2019). "On Aerts' overlooked solution to the EPR paradox." In: Probing the Meaning of Quantum Mechanics –

Aerts' reasoning was as follows. In their premise, EPR had assumed that for two quantum entities, like two electrons, a spatial separation was equivalent to an experimental separation. Moreover, they had assumed that the quantum formalism was able to correctly describe such a situation. In other words, they implicitly hypothesized that quantum theory could describe a system formed by two *experimentally separated* physical entities. But since this has generated a contradiction, such assumption is in fact incorrect, namely:

> *Quantum theory cannot describe physical entities that are experimentally separated.*

It could be argued that the mistake made by EPR was simply to believe that a sufficient spatial separation between two electrons (or between two other microscopic entities) also necessarily implied their complete *disconnection*. On the other hand, the experimental physicists, with their very sophisticated experiments, have shown that if you take sufficient precautions it is possible to create experimental conditions where two microscopic entities, like two electrons, after having interacted, always remain *interconnected*, even when separated by arbitrarily large spatial distances.

The real and only error of EPR was therefore to have applied their reasoning to a wrong experimental situation, that is, to the situation of a bipartite system formed by two entangled entities, because when two entities are in that particular "state of intricacy," the experiments carried out by Alice and Bob can only give perfectly correlated results, as the numerous experiments afterwards confirmed.

In their time, EPR could not of course know that the notion of entanglement, which appeared at a mathematical level in the theory, was also an expression of a real phenomenon, so they could not suspect that they were making such an error. But their logical reasoning was nevertheless correct, and it has never been invalidated by experiments (no logical reasoning can be

Information, Contextuality, Relationalism and Entanglement, World Scientific, pp. 185–201.

invalidated by experiments, only its premises can be).

Imagine a situation in which the experimental physicists, instead of taking every precaution to maintain the interconnection between the two electrons, perform what would be interpreted as "a badly executed experiment," that is, an experiment where the focus is not in highlighting the presence of correlations, but instead in trying to highlight an absence of correlations.

We remind you that in the experiments conducted over the years what experimenters only ever tried to do, by any means, is to produce a violation of Bell's inequalities, that is, everything has been always put in place to preserve the so-called *coherence* between the two electrons (or photons, or other microscopic entities). But it is also possible to imagine experimental situations (never actually explored to be honest, not deliberately at least) where one would instead do everything to highlight an absence of coherence between the two electrons, i.e., a condition of *disintricacy*, or *disentanglement*.

The realization of such experimental situations would highlight a separation between the measurements of Alice and Bob, leading exactly to the contradiction that was pointed out by EPR. Situations of this kind are in fact perfectly common when we are dealing with non-microscopic physical entities. In other words, the reasoning of EPR makes it possible to conclude that quantum theory is not able to describe experimental situations where the properties that are measured, relative to two distinct physical entities, would remain uncorrelated. In other words, situations where the two entities would be (not only spatially but also experimentally) separate.

Clearly, if our physical reality is "an interconnected whole," that is, everything is in a permanent state of intricacy with everything else, this structural shortcoming of quantum theory, in describing separate entities, would not be such, but rather a correct description of how things would really be at a fundamental level.

On the other hand, we live surrounded by macroscopic entities that, apparently, and until proven otherwise, show no particular quantum effects, and it is unclear whether it will ever be possible, for example, to put two chairs in our living room into a *bona fide* quantum state of entanglement.

So, we can say that the question of whether the structural insufficiencies of the quantum formalism highlighted by Aerts are a serious problem or not, for a theory that aims to describe our physical reality at every possible level, remains today still open.[52]

All right, we have thus clarified that quantum theory is not necessarily a complete theory, as it adequately describes only the situations where the systems are always in a possible state of interconnection, while it is legitimate to suppose that there are portions of reality that are able to remain disconnected one from the other, and that a complete physical theory should be able to describe both these situations, of connection and disconnection, as well as all the possible intermediary situations, but to do this a structurally richer formalism is needed, which goes beyond the standard quantum formalism (in this case, we speak of *non-Hilbertian* theories).

That said, the important and surprising fact remains that, through the violation of Bell's inequalities, the existence of correlations of the second kind, among microscopic entities separated by arbitrarily large spatial distances, has been evidenced in the structure of our physical reality. A question then arises:

---

[52] For the "insider" reader, it should be pointed out that the significant part of Aerts' work has been to highlight, in a constructive way and not through an *ex-absurdum* reasoning, what are the "missing elements of reality" of quantum theory. These missing elements do not manifest themselves at the level of the "state space." In fact, quantum theory has a kind of overabundance of states, due to the so-called *superposition principle*. The problem instead manifests itself at the level of the properties, which in the quantum formalism are described by particular operators, called *orthogonal projections*. It is precisely the overabundance of states that produces a corresponding underabundance of properties, in the sense that certain properties that are characteristic of bipartite systems formed by (experimentally) separate components cannot be represented by using orthogonal projection operators. Therefore, if these properties are considered to be actualizable, quantum theory would be incomplete because unable to represent them.

> *How can two entities spatially separated by possibly astronomical distances remain nevertheless interconnected?*

The answer is simple: they can do it because they are *non-spatial* entities, that is, entities that when they are in a state of entanglement form a "non-spatial whole." In other words, their connection remains invisible to us because it is a connection that does not take place through space.

When Alice and Bob jointly measure the properties of two entangled electrons, which are only apparently separated, what actually happens is that a *unitary non-spatial entity* is separated into two parts, just like when we pull an elastic and break it, so creating two separate fragments of elastic, whose lengths will necessarily be perfectly correlated.

Try to imagine Alice holding in her hand one end of a very long elastic band, while Bob, at a considerable distance from Alice, holds the other end (see Figure 30). Imagine that Alice and Bob have agreed to pull hard on the elastic in a predetermined instant, causing it to break.

When Alice (respectively, Bob) receives in her (his) hand her (his) own fragment of elastic, she (he) can measure its length, and knowing the original length of the elastic, can deduce the length of the fragment received by Bob (respectively, by Alice), without having ever communicated with her (his) colleague (see Figure 30).

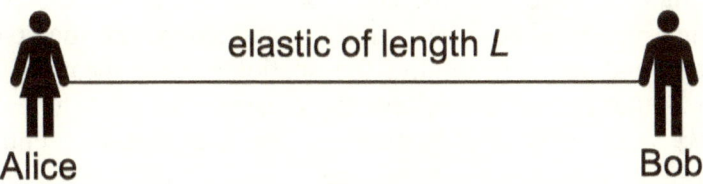

Figure 30. *Alice and Bob both hold one of the two ends of an elastic of length L.*

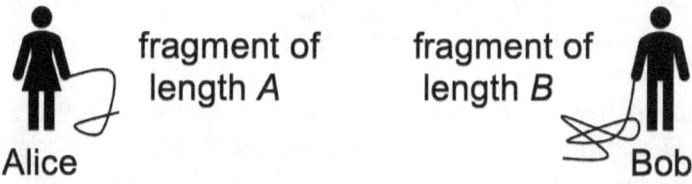

Figure 31. *If Alice's elastic fragment is of length A, that of Bob is necessarily of length B = L - A. In other words, the lengths of the two fragments are not arbitrary but perfectly correlated with each other.*

This is exactly what happens in the quantum laboratories, when bipartite systems in entangled states violate Bell's inequalities, revealing the presence of correlations of the second kind. We can observe that Alice and Bob jointly operate on a unitary entity: the intact elastic. Nevertheless, you can conceive of this elastic as a bipartite system, as it has two well-defined and distinct ends, which Alice and Bob can grasp with their hands.

These ends of the elastic are connected to each other by the structure itself of the elastic, which is entirely present in space, therefore perfectly visible, whereas in the case of microscopic bipartite entities such a connection would be invisible, as non-spatial in nature.

Einstein did not accept the idea of quantum entanglement essentially for two reasons. The first is that he believed that the physical reality should be entirely contained within the spatial theater (and more generally, the spatiotemporal theater). The other is that he did not believe the existence of what he called *spooky actions at a distance*, i.e., the fact that two entities separated by arbitrarily large spatial distances could communicate with each other at virtually infinite speed, far superior to the speed of light, which according to his *relativity theory* was an impassable speed limit for each signal capable of producing an effect.

But the prejudice that physics must portray a reality "wholly contained in space" cannot be considered an *a priori* principle, and it is possible that in our day Einstein would have surrendered to the evidence that our physical reality is mainly non-spatial in nature.[53] In doing so, he would also automatically have solved his perplexity about the "spooky actions at a distance," since the very notion of "action at a distance" does not apply to non-spatial entities.

Moreover, if we consider the paradigm of the experiment with the elastic band, it is clear that during the process of creation of correlations, Alice and Bob do not send any signal to each other, but simply act in concert on the same unitary entity.

Now, in Chapter 11 we have underlined how non-spatiality can be an expression of the fact that the microscopic physical entities would behave in a way that is strikingly similar to how conceptual entities behave. According to this conceptual perspective, the non-spatial connections subtended by quantum entanglement would be nothing more than *connections through meaning*, i.e., connections that would result from the fact that concepts are "entities of meaning" and that what measures concepts, the cognitive systems, are entities that are sensitive to their meaning.

So, if for example you ask a person to give an example of an "animal eating food," for that person's mind the two concepts "animal" and "food" will appear to be mutually entangled, since they are connected through the meaning conveyed by the sentence "animal eating food."

This can be verified by noting that if for example the animal chosen is "horse," the corresponding food that will be chosen cannot be any food, but most likely (although not necessarily) will belong to those foods that are considered a horse can eat, such as "hay," "oats," "apples," etc.

In other words, if the invitation to choose an example of "animal eating food" is considered as a measurement jointly

---

[53] From the perspective of the theory of special relativity, non-spatiality emerges, for example, in the disconcerting observation (disconcerting if we limit ourselves to a purely spatial view) that the speed of light is exactly the same in every possible inertial frame of reference. See: Aerts, D. (2018). Relativity Theory Refounded. Foundations of Science 23, pp. 511–547; doi: 10.1007/s10699-017- 9538-7.

carried out on the "animal" entity and on the "food" entity, to be repeated innumerable times (with different people), it will be noted that certain combinations, like "horse eating hay," will be more frequently selected than other combinations, like "horse eating steaks," or "cat eating hay."

So, the different choices of examples for "animal eating food" will necessarily create correlations between the "animal" and "food" exemplars, and these *correlations of the second kind*, like those generated by the microscopic entities (which would be also of a conceptual nature), result from the presence of an evident, though abstract, "connection through meaning."

Therefore, the quantum states of entanglement, and the connections through meaning of the conceptual entities, would be only two substantially equivalent ways of talking about the same phenomenon in nature, of a genuine non-spatial nature.[54]

---

[54] It is interesting to note that it is possible to conduct psychology experiments that exploit precisely such "connections through meaning" between different concepts, when in certain states, to violate Bell's famous inequalities. See for example: Aerts, D. & Sozzo, S. (2014). Quantum entanglement in conceptual combinations. International Journal of Theoretical Physics 53, pp. 3587–360. Aerts, D., Aerts Arguëlles, J., Beltran, L., Geriente, S., Sassoli de Bianchi, M., Sozzo, S & Veloz, T. (2018). Spin and wind directions I: Identifying entanglement in nature and cognition. Foundations of Science 23, pp. 323–335. Spin and wind directions II: A Bell State quantum model. Foundations of Science 23, pp. 337–365.

13. OTHER "OBSERVER EFFECTS"

This chapter aims to provide a quick overview on the concept of "observer effect" outside the field of physics.[55] The term "observer effect" usually refers to the possibility that an observation can affect or even create the properties of what is observed. However, depending on the context and the mechanisms at play, it may indicate effects of a very different nature.

On the quantum observer effect, you already know a lot, but the term is more generally used even in situations where a measurement cannot be considered to deliver a perfectly exact outcome, because the method used can alter in part the result.

A typical example is when we measure the pressure of a tire and let some air out when we insert the pressure gauge, or when we measure the temperature of a liquid and the difference in temperature between the thermometer used and the liquid itself alters the temperature of the latter. In this type of situations, we usually speak of *probe effects*.

In the field of programming, we talk instead of a *heisenbug* (the term is a pun referring to the name of the physicist *Werner Heisenberg*), to indicate a bug in software that is able to alter its behavior, or even disappear, when one tries to probe it.

In the social sciences, the American linguist *Amber Labov* has instead introduced the term *observer's paradox*, to describe those situations in which the presence of an observer is able to alter the outcomes of an observation. In sociolinguistics, for example, when a researcher tries to collect data on the use of natural language, by interviewing people, with her/his presence alone s/he can induce a change in the way they will speak,

---

[55] See also the entry "Observer Effect", that I had the pleasure of taking care of in the encyclopedic volume: *The SAGE Encyclopedia of Educational Research, Measurement, and Evaluation*, edita da Bruce B. Frey, Thousand Oaks, CA: SAGE Publications (2018).

because the very context of the interview can induce them to speak in a more formal way than usual, and therefore no longer according to a genuine natural language.

More generally, the observer's paradox, also called *Hawthorne effect*, describes those situations in which the behavior of people is changed in ways that can hardly be foreseen by the experimenters, simply because they are monitored or inserted in a given experimental context.

As for the specificity of the observe effect of a quantum kind, as we already mentioned, quantum formalism has been successfully applied in the modeling of human decision-making processes, within that new field of research called *quantum cognition.*

The reasons for this success are numerous, but one of them is really linked to the quantum observer effect, which has its natural counterpart in psychology. In fact, in many interrogative contexts, the answers that are obtained, when people are submitted to a questionnaire, are not only discovered, but often they are literally created, in a completely unpredictable way.

For example, consider a survey that asks 100 people whether they are smokers or non-smokers. If 50 have answered affirmatively and 50 negatively, we can state that the probability of finding a smoker in the group of participants, choosing it randomly, is 50%, just like in the example of the box containing the 100 white and black assorted elastics of Figure 4. Evidently, this probability reflects the actual behavior of the 100 people in question, regarding their way of relating to smoking.

But suppose now that those same people are asked whether they are in favor or against the use of nuclear energy. Imagine that even in this case 50 of them answer yes and 50 answer no. Once again, we can say that we have 50% probability for one type of answer and 50% probability for the other type of answer. But can we interpret these probabilities by believing that, before the question had been asked, 50 people in that group were in favor of nuclear energy and 50 were against it?

This interpretation would clearly be wrong, since people with a well-defined opinion about the nuclear issues are rare, which means that most will be forced to actualize an answer at the moment, in a perfectly indeterministic way. We therefore find ourselves in a situation that is very similar to that of the left-handedness test of Figures 13-15. In other words, this time the answers are not simply discovered, but literally created, in a way that does not depend solely on the state of the participants and how the question is formulated, but also on the unpredictable fluctuations that manifest in their mind when confronted with that specific cognitive situation.

So, if a survey like the one we have just described is interpreted as a measurement process, we can say that we are in the presence of an observer effect, because the process is undoubtedly invasive (the participants are somehow forced to give an answer, so to "break the symmetry" of their possible doubt) and is able to create those same properties that are observed.

Another important example of an observer effect is the fact that some observations may disturb one another and are therefore experimentally incompatible with each other (as expressed in Heisenberg's famous uncertainty principle). This means that if we sequentially execute two measurements that are not mutually compatible, the order of the sequence will have an influence on the statistics of the results obtained. Measuring the position first and then the momentum of an electron is not the same as measuring the momentum first and then the position. The same happens in the psychological ambit.

When we ask a sequence of questions, their order can influence the answers that are provided. For example, asking first, "Is Bill Clinton honest?" and subsequently "Is Al Gore honest?" does not produce the same statistics of answers than asking first "Is Al Gore honest?" and then "Is Bill Clinton honest?". These *order effects* are obviously a source of concern for psychologists and sociologists, when they study people's beliefs, attitudes, intentions and behaviors, and a stratagem to mitigate these "observer effects" is to always randomize the orders of the questions, so that the respondents do not always answer according to the same sequence.

We can also mention the famous *quantum Zeno effect* (the name derives from the famous paradox of the arrow devised by the Greek philosopher *Zeno of Elea*), a situation in which the continuous observation of a system can "freeze" its evolution.

For example, if an unstable atom is observed very frequently, its decay can be prevented.

A similar effect has also been described in the field of neuroscience, noting that a continuous focus of attention is able to stabilize the neuronal circuits of the brain. In a completely different field, the effect produced by multiple observations is also described in the psychological phenomenon known as the *bystander effect* (also called the *bystander apathy*), according to which the more spectators are present in an emergency situation and less likely one of them will intervene to provide help.

To conclude, we would like to mention a last circumstance, also often described as an observer effect, capable of affecting the collection and analysis of data and the design of research. It happens when the desire to observe something is so strong that it causes people to believe what they want to believe, that is, to "observe" something that does not really exist.

An emblematic example is that of the famous *N-rays*, whose "discovery" took place in 1903, by the French physicist *René Blondlot*, followed by numerous studies and publications confirming their existence, by more than a hundred respectable scientists, in about 300 scientific articles published in prestigious journals. Instead, it was an exemplary case of an entire scientific community mistaking candles for lanterns, deceiving themselves for many years, probably also due to the recent discovery of *X-rays* and the strong expectation about the possibility of easily discovering new forms of radiation.

The scientific method has been designed precisely with the aim of neutralizing our errors of assessment, our false expectations, our prejudices, and other mechanisms of self-deception, but, of course, our state of alert must always remain high, as our journey of progression in knowledge does not guarantee us in any way to avoid the pitfalls of our prejudices, not only individual but also collective.

AFTERWORD

I remember reading my first popular book of science around the age of fourteen, borrowing it from my cousin. It was the book "*Dagli atomi al cosmo*" (From atoms to the cosmos) by *Piero Bianucci*. I also remember that shortly after I saw in the window of a bookstore a booklet with a sky-blue cover, titled "*L'universo di Einstein*" (Einstein's Universe) by *Nigel Calder*,[56] which I immediately bought and read avidly.

If the book of Bianucci opened my young mind to the majestic sceneries of the cosmic dynamics, by telling me about stellar evolution, blacks holes, and possible messages coming from extrasystemic civilizations, the booklet of Calder worked in me more deeply, revealing me how a well-educated human mind – in this case that of the great Einstein – was able to unravel the profound mysteries of the reality in which we live, bypassing our more radical (and deep-rooted) prejudices on how we believe things should be, although in fact they are not.

Those readings awakened something important in me: the memory of a possibility that with time I would have learned to recognize and explore more in depth. So it was that when I obtained the *diplôme de bachelier* (the French school-leaving certificate), I decided to enroll in the faculty of physics at the *University of Lausanne* (Switzerland), not without taking my father by surprise, as he had envisioned for me a career in management, as was the tradition in the father's side of the family.

Despite a somewhat uncertain start (in those days I was more interested in the lightness of the student life than in books), I succeeded in my attempt to complete my studies, without great difficulties to tell the truth, also because, contrary to what is usually believed, studying physics (or mathematics) is

---

[56] The booklet was written in *1979*, on the occasion of the centenary of the birth of Einstein, and was published in Italy two years later.

quantitatively much less challenging than studying subjects such as law or medicine, as although there are many more things to *understand*, there are also far fewer things to *memorize*.

In this way, with a somewhat minimalist style, I came to earn my degree in *1989*. In this first part of my scientific journey, I felt out of love – so to speak – about topics such as astronomy and astrophysics, not because of their content, but because of the lack of charisma and the poor teaching skills of those who in those times, and places, taught them. Fortunately, in exchange, I found some very good teachers that were able to infect me with their passion for more specifically theoretical subjects, such as quantum physics and relativity. Thus, ultimately it is the booklet of Calder which prevailed, not the one of Bianucci.

So, I decided to choose the direction of becoming a researcher in theoretical physics, also because, somehow as it happened to Pauli, when I tried to carry out experiments in the laboratory, equipments often broke down, or began to operate in a totally anomalous way, especially if they were instruments of an electronic nature.

For those who have never heard of the *Pauli effect*, let me remind you that the great Austrian physicist was known for his talent in compromising the outcome of any experiment in physics, by means of its mere presence. So much so that his experimentalist colleague and friend *Otto Stern*, ended up by categorically prohibiting him to access his laboratory.

The Pauli effect, incidentally, brings us back to the theme of this booklet, as it is, or rather it would be, a kind of "observer effect," related to the possibility of so-called *macro-psychokinesis* phenomena. It must be said that Pauli did not think to the effect bearing his name as a mere playful way of describing a series of unfortunate events, hypothetically related to the presence of his person, but as an effect perfectly objective (Pauli, it is good to mention, was a firm supporter of research in parapsychology).

Anyhow, regardless of my ability, real or pretended, to create anomalies in the electronic equipments, my interest was more directed at the theoretical, rather than the experimental, investigation (although, of course, my talents as a theoretical physicist were not in any way comparable to those of Pauli, or of giants of his stature). So, after graduation I was pleasantly surprised to see me offering an assistant position at the famous *School of Physics* of the *Geneva University*, with the possibility of undertaking a research project aimed at obtaining a doctorate.

It was then that I got in touch with *Constantin Piron*, one of the co-founders of the *Geneva-Brussel* school, which I have frequently mentioned in this book. At that time, mainly because of my scientific immaturity, I had no way of fully appreciating the depth of the conceptual ideas of Constantin, with whom I took however to work, in view of a possible PhD thesis.

The work he proposed me to carry forward under his direction was about a possible reconstruction of *quantum electrodynamics* (the relativistic quantum theory of the electromagnetic field) starting from a critical view of the role of the *observer in relativity theory* (expressed by some rather original ideas of his about a proper interpretation of the so-called relativistic covariance).

It is interesting to note that, already at that time, unbeknownst to me, I was confronting myself with the analysis of the role of the observer in the description of the physical reality. I say "unbeknownst to me" because in the end my scientific collaboration with Constantin did not produce the desired outcome, also because in those days I was still somewhat confused and uncertain, as well as skeptical, about the content of his ideas, certainly innovative but at the same time rather controversial.

Undoubtedly however (especially as I was able to realize many years later), having been his assistant for more than a year, interacting with him almost on a daily basis (especially in the many bistros of the Geneva area!), irreversibly changed my way of looking at the mysteries of the quantum world, which, somehow paradoxically, used to simultaneously become less and more mysterious when I was in his presence.

After my visit to Geneva, I returned to Lausanne, this time to the *Ecole Polytechnique Fédérale* (*EPFL*), where I began a fruitful collaboration with *Philippe A. Martin*, who like Constantin was a student of the Swiss physicist *Josef-Maria Jauch* (who in turn has been the assistant of Pauli), famous not only for his work in the mathematical and epistemological foundations of quantum mechanics,[57] but also for being one of the founders of modern *quantum scattering theory*.

In the ambit of this theory, which describes the collision processes of the different microscopic entities, a particularly delicate problem was the study of the *time-delays* induced by the different interactions. The problem was tricky because, even though it was possible to describe unambiguously in quantum physics the probability of detecting a microscopic entity at a given point *x*, at a given instant *t*, conversely it wasn't possible to calculate the inverse probability of arriving at a given instant *t*, at a given point *x*. In other words, the concept of *time of arrival* was not definable in the theory, given that surprisingly it didn't contemplate the possibility of a *time observable*.

It was the already several times mentioned Pauli, by means of a famous *ad absurdum* reasoning, who demonstrated that if one assumes the existence of a time observable in quantum theory, then one arrives to an inevitable contradiction.[58] On the other hand, there was a simple, but ingenious way to bypass the problem, which was to replace the concept of *arrival time* with the similar, but not equivalent, concept of *sojourn time*. The latter, contrary to the former, was definable in a unique way in quantum theory, and therefore there was an appropriate path to study time-delays and their properties in quantum mechanics (then defined as a difference of two sojourn times instead of a difference of two arrival times).

It was on these kinds of problems that I began to work with Philippe, when I started collaborating with him at the former *Institute of Theoretical Physics* of the *EPFL*. This time the collaboration was successful, and I had the satisfaction of writing and publishing a number of research papers in international journals, which went on forming the core of my doctoral thesis, which I ultimately defended in *1995*.

---

[57] See for instance his delightful booklet: *Are Quanta Real? A Galilean Dialogue*, Indiana University Press.

[58] See for instance the discussion in: Sassoli de Bianchi, M., *Time-delay of classical and quantum scattering processes: a conceptual overview and a general definition*. Central European Journal of Physics, Volume 10, Number 2, Pages 282-319 (2012).

The doctorate is the transition to adulthood for an academic researcher. In fact, as is written on the same certificate: *by acquiring the degree of doctor the candidate demonstrates his aptitude for scientific research*. The "demonstration" obviously occurs through the exposure to the constructive criticism of other researchers (the so-called "peers"), as is customary in a science worthy of the name. So, not without a certain satisfaction, I became a theoretical physicist "with certificate of adulthood." In those years however, I had to face other problems associated with adulthood, of a very different kind than quantum scattering processes.

I was in fact married and already father of two children (plus two dogs and a cat), and a number of responsibilities, particularly economic, burdened on my shoulders. After the PhD I thus found myself at a turning point: either I continued my research in the academic field, accepting the rules of the game (the wages of post-doctoral students, especially outside of Switzerland, were not particularly attractive) or, simply, I abandoned scientific research in view of a more lucrative job. And since this more gainful job was offered to me on a silver platter, it seemed reasonable to me, at that time, to choose the second option.

This choice, however reasonable, produced in me an uncomfortable cognitive dissonance, because I always felt as being a researcher in the soul. To solve this inner conflict, I promised myself I would find some way to continue to do research, although not anymore in a strict academic context. For a while (about 3 years), I succeeded in the intent of earning money during the day and solve physics problems during the night, cultivating a few scientific collaborations at a distance. But considering family commitments, and the limited capacity of my liver in assimilating massive amounts of coffee, I soon arrived at a breaking point. Also because, outside of the space-times of a research environment, it becomes very difficult to keep alive not only the interest, but also the ability and the pleasure in producing quality research, failing that daily confrontation of ideas which is at the basis of any investigation.

But to the extent that, willy-nilly, I abandoned the research in physics, I also started to intensify and deepen a different approach to research, toward which I had always felt drawn: *inner research*. Now, although there are considerable differences between the content and methodology of an exterior and interior research, there are also many points of contact, and certainly some profound analogies. In fact, both a physicist who explores the mysteries of the cosmos, and a self-researcher investigating the so-called "spiritual dimensions," the observations occur in areas normally inaccessible to the ordinary senses: for modern physics they are typically the microscopic and astronomical dimensions, while for the self-researcher they are the inner dimensions, accessible only through non-ordinary states of consciousness.

Both these approaches require, in addition to specific (external or internal) technologies to allow one to experience these hidden levels of reality, also a suitable language, sufficiently evolved, to be used to effectively describe the results of these observations-experimentations. It is mainly in this ambit that the research of a theoretical physicist, who is interested in the foundations of physical theories, can find possible synergies with inner research. And it is during my search for such a possible language that, quite unexpectedly, I found myself interested again in the pioneering work done by the founders of the Geneva-Brussel school. More precisely, I became interested in the writings of *Diederik Aerts*, a student of Constantin, who unlike him possessed, in addition to the gift of clarity, a remarkable talent for the *transdisciplinary* visions.[59]

Thus, I started to integrate the suggestive ideas of Diederik in some of my writings about self-research. At that time I also decided to abandon my career as a manager and for a while I returned to teach physics, this time at the high school of Lugano. I also founded a small private laboratory, aimed at the study and teaching of self-research,[60] and in that context started again, to my great surprise, to get actively involved in physics. I began by writing some popular texts, and then I had the

---

[59] Transdisciplinarity is an intellectual and scientific approach that seeks a full understanding of the complexity of reality.

[60] This is the *LAB – Laboratorio di Autoricerca di Base*; see: *www.autoricerca.ch*.

pleasure to get in touch with Diederik, who encouraged me to continue my effort of clarification and dissemination of the approach that he and Constantin had originated, and that he with his group in Brussels were further expanding.

So, subsequent to my "turning point" at the end of the PhD, and although it had been nearly two decades, I started once more getting professionally involved in fundamental research in physics, writing and publishing again research papers, although this time as an independent researcher, and having also collected along the way the precious instrument of self-search, which remains today my major field of interest and activity.

This double role of mine, both as a physicist, in the most traditional sense of the term, and as a self-researcher, makes me a person with some distinctive characteristics. Let's be clear, I'm certainly not the only person in the world being conjunctly interested in physics and spirituality; on the other hand, there are not so many individuals that promote both a pragmatic and disenchanted approach to the inner research, not tainted by unnecessary religious dogmatism or cultural bias (as far as possible), and a thorough research on the foundations of physical theories.

There are many self-proclaimed spiritual leaders who today speak and write – often inappropriately – about quantum physics, posing as the real experts in the field, when at best they have read a few popular booklets, or watched some videos on the Internet. On the other hand, it is well-known that it is sufficient nowadays to add the word "quantum" to the description of an experiential workshop, to immediately increase the number of persons enrolled, or to the title of a book of spirituality to consistently increase sales. Man has always been looking for certainties, and quantum physics, as a science, seems to have become the new tool to be manipulated in order to achieve, whatever the cost, such certainties.

Of course, as scientific culture is often scarce in the ambit of spiritual research, especially in so-called *New-Age* movements, in the same way a correct understanding of what are the contents of a serious spiritual research is also absent in many institutional scientists, including physicists. This is so because not only do they lack sufficient experience in the field of non-ordinary perceptions, but also because they adhere, without knowing it, to the vision of a strict *physicalism*. I say "without knowing it" not because they wouldn't be aware of supporting such a philosophical position, which considers that all knowledge, in ultimate analysis, can be traced back to the statements of physics, but because they are often not aware that the rigidity of such a position has often the flavor of a true dogma, which unnecessarily restricts the explanatory power of their theories of reality.

Curiously, we find traces of that same physicalism also in the attempt of many "gurus" of our days in giving a foundation to the different paranormal phenomenologies (the famous *spiritual powers* – or *siddhis* – as described for example in the ancient path of *Yoga*) by means of quantum physics, forcing its interpretation in order to "scientifically prove" the action of the observing mind over matter.

Of course, anyone who has had sufficiently significant personal experiences in the ambit of parapsychism will be motivated by the desire of finding a legitimate explanation for such experiences, that is, some mechanisms able to explain phenomena such as telepathy, clairvoyance, precognition and premonitions, travelling clairvoyance and extracorporeal states, and so on. And of course, the idea of a possible psychophysical dynamics associated with the quantum observational process seems to come in handy!

It must be said that the founding fathers of quantum physics were very interested in issues of spiritual research, and carried on a rather substantial and prolonged debate in the attempt to determine what could be, if any, the mystico-metaphysical implications of the new theory,[61] especially with regard to a possible action mechanism of the mind over matter.

But today we can affirm, based on a more mature and disenchanted vision of quantum theory, that in no way it contemplates such a mechanism. This is because, as I tried to explain in this booklet, it is absolutely not necessary to call upon it to understand the origin of quantum probabilities; and of

---

[61] Marin, J. M., "'Mysticism' in quantum mechanics: the forgotten controversy." Eur. J. Phys. 30 (2009), pp. 807–822.

course, it makes no sense to introduce additional *ad hoc* explanations when they do not correspond to a specific *cognitive gap* (*Occam*'s razor).

This is to say that despite having personally experienced many of the so-called paranormal phenomena, I can assert with full knowledge of the facts – having "a foot in both camps" – that it is totally illegitimate to exploit the quantum observer effect in order to try, at all costs, to give a scientific basis to the action of the mind over the matter-energy. From my point of view, wanting to do so is either being a bit naïve in scientific terms, or intellectually dishonest. By the way, paradoxically, insisting in doing so also means to adhere, as already stated, to a strict physicalism, namely to a philosophical view that is usually closed with respect to the spiritual dimensions.

Let me explain better, and on this I will end my personal note. Physicalism believes that our scientific explanations are based on the knowledge of the laws of physics. But let us not forget that physics, in turn, basically founds its knowledge on the collected experimental data. These data are accumulated by making use of special instruments of observation and measurement, which are generally constituted by ordinary macroscopic objects, of the *inanimate* kind.

One of the implicit assumptions in the vision of physicalism is that inanimate bodies are essentially equivalent, from the point of view of their basic properties, to *living* bodies, in agreement with the point of view of the modern neurosciences, who consider consciousness a mere *epiphenomenon.*

On the other hand, if we take in due account the numerous intersubjective data collected by self-researchers of any geographical region, culture, age and gender, in the different epochs, who have been involved in the exploration of so-called "subtle dimensions" of life, it is possible to assume (even though only speculatively) that living organisms, and especially humans, possess an "expanded physicality." Namely, that their field of manifestation would go beyond what our laboratory inanimate instruments would be able to detect. If this hypothesis is acceptable, as I think it is, there would be no meaning in looking for possible psychophysical mechanisms in the ambit of quantum physics, since this theory, as advanced and sophisticated it may be, has never dealt with subtle matter fields, associated with the living, but only with ordinary matter fields, associated with ordinary inanimate instruments.

There is no doubt that reality is not independent from the participatory minds (or consciousnesses) that populate it. Minds, as is known, can act in reality by means of the bodies through which they manifest, carrying out actions that can promote both discoveries and creations.[62] Every time we drink a simple glass of water, our mind acts within reality, which therefore cannot be considered totally independent from it.

The relevant scientific question, which is still open, is not therefore to understand whether or not a mind is able to act, or interact, over matter and energy, since we already know it can do it, but rather to determine in how many different ways it can do it, i.e., which, and how many, would be the vehicles of manifestation of a human consciousness, in the different levels and planes of the vast multidimensional reality.

This question opens up to future scenarios of research (for some readers certainly of a sci-fi kind), where the conventional research in physics, which only uses inanimate instruments, could be integrated in an ampler form of investigation, also employing living equipments (such as human beings properly formed and trained for this purpose), able to detect the entire spectrum of the multimaterial fields that possibly characterize the dimension of life, and therefore of reality, about which humans seem to have experiences since the dawn of time.

I conclude by leaving the word to the American physicist *Harold Puthoff*.[63]

---

[62] This statement is not inconsistent with the hypothesis of realism. It is important to distinguish a physico-energetic process, by means of which a mind acts in reality, changing it, from a mere conscious representation of the various phenomena a mind can have access to, through its experience.

[63] Taken from a discussion held during the *Colloque de Cordoue*, organized in 1979 by *France-Culture*, entitled: *Science and Consciousness: the two readings of the universe.*

When one of our subjects completes an experiment in psychokinesis, we always ask him: "How you did it? What happened within yourself?" And the answer that is always given to us [...] is that the only thing that the subject has done, was in a certain way to find where there was life in the object, and that the object had momentarily become something alive.

autoricerca.com

HEISENBERG UNCERTAINTY PRINCIPLE AND THE PHYSICS OF SPAGHETTI

Massimiliano Sassoli de Bianchi

ABSTRACT. This is an article written in a popular science style, in which I will explain: (1) the famous Heisenberg uncertainty principle, expressing the experimental incompatibility of certain properties of micro- physical entities; (2) the Compton effect, describing the interaction of an electromagnetic wave with a particle; (3) the reasons of Bohr's complementarity principle, which will be understood as a principle of incompatibility; (4) the Einstein, Podolski and Rosen reality (or existence) criterion, and its subsequent revisitation by Piron and Aerts; (4) the mysterious non-spatiality of the quantum entities of a microscopic nature, usually referred to as non-locality. This didactical text requires no particular technical knowledge to be read and understood, although the reader will have to do her/his part, as conceptually speaking the discussion can become at times a little subtle. The text has been written having in mind one of the objectives of the Center Leo Apostel for Interdisciplinary Studies (CLEA): that of a broad dissemination of scientific knowledge. However, as it also presents notions that are not generally well-known, or well-understood, among professional physicists, its reading may also be beneficial to them.

1. INTRODUCTION

This article contains the (revised and slightly expanded) "transliteration" of a video that I published on YouTube on April 5, 2012, first in Italian [1] (my native language), then on August 27, 2012 also in English [2]. The Italian version received to date (June 15, 2018) more than 148,000 views, and considering that it is a video of almost an hour and a half, which is exclusively about physics, I consider it as an encouraging result. This also explains why at the time I decided to make the effort of producing an additional English version of the video, which however, probably due to my "macaronic English," only obtained to date a little more than 18,000 views: a one order of magnitude difference with respect to the original Italian version.

Regardless of the differences in terms of numbers of views, both the Italian and English videos received very positive comments (an event in itself quite rare on YouTube), which is the reason that led me to also write the present article. I hope in this way to do something pleasing to those who enjoyed the video, offering them the opportunity to retrace its contents in a form not only stylistically a bit more accurate, but also, perhaps, more suitable for the continuation of the reflection about its content. I also hope that this will allow the non-habitual users of YouTube to also access the explanations contained in the video, and that the present article version of it will be met with the same enthusiasm.[1]

Before starting, let me bring back some of the positive comments I received in connection with the video. This not to indulge in some kind of narcissistic pleasure, but because these extemporaneous comments (here taken from the Italian version of the video) are able to express, I believe, some of the characteristics of the text that I hope you are in the process to read.

"Interesting and really well done. I deal with the philosophy of science in the USA; just a few hours ago I debated on the linguistic difficulty related to the sayability of the concepts of quantum mechanics in an unambiguous way and different from the classic

---

[1] A preprint version of the article is also available on the archive (*arXiv.org*) of Cornell Univesity: *https://arxiv.org/abs/1712.08465.*

ontology, which generates misunderstandings and often inaccuracies; I will report your valid presentation as an excellent example. Congratulations."

"Thank you for the explanation [...] I also liked the part about the spaghetti, a brilliant metaphor to understand the influence of the experimenter on the physical system, far beyond the banality of classical concepts."

"I find the explanation incredibly clear for those who want to have a general idea of the problems. The uncertainty principle drove me crazy because of its incompatibility with everyday experiences and it cut me off from a somewhat deeper reading on quantum physics. Thank you for giving us your time and your expertise!"

"Really interesting. Thank you. But now I hate wooden cubes to death..."

Of course, there were also some less enthusiastic comments, more critical about the content of the presentation and the way things were explained. Here is an example:

"Despite having a great (amateur) passion for the topic, after six minutes I got bored and lost... if you dive into charts and formulas, then the video is not for everyone. [...] If the things explained in the video I had read them in a book, it would have been the same."

Well, I hope that this last comment, although not very laudatory, portends a possible success also for the "article format" of my video-work.

To conclude this brief introduction, I would like to say that most of what I will tell you in the following pages, is the result of an understanding that has developed thanks to the work of the so-called Geneva-Brussels school of quantum mechanics, especially thanks to the research of Constantin Piron (of whom I was the assistant in Geneva, for his famous course in quantum mechanics) and Diederik Aerts (a student of Piron, with whom I have the pleasure today to collaborate).

Very well, I hope your reading will be enjoyable and thought provoking.

2. A SIMPLE EXPERIMENT

Let me begin with a very simple experiment. We are on the surface of a frozen lake, by night. The physical system that we want to study is a wooden cube, and the instrument at our disposal to do so is a camera with flash. The experimental procedure is as follows: a colleague throws the cube on the ice, so that it will slide on the surface of the lake (without letting it rotate and assuming for simplicity that there is no friction).

At this point, we take a first picture, at time $t = 0\ s$. This first picture shows us that the wooden cube was, at that precise moment, in the position x_0 (see Figure 1).

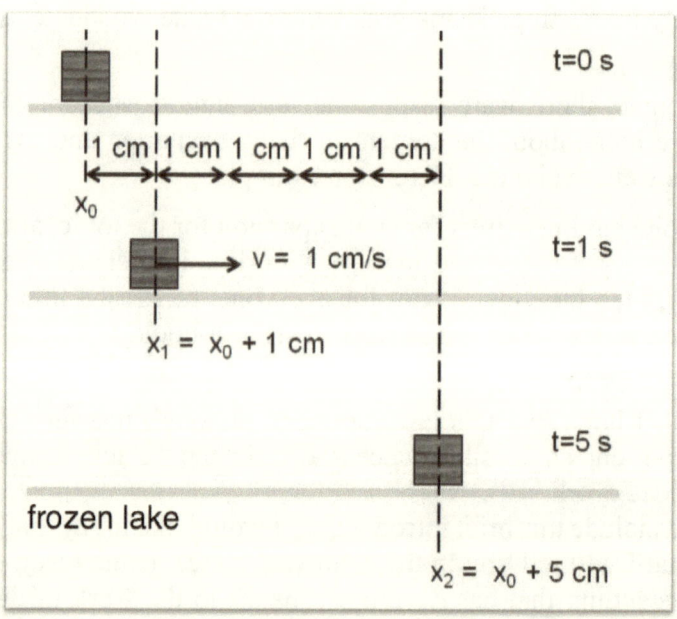

Figure 1. *The three snapshots indicating the different positions of the wooden cube, at the three different instants of time $t = 0, 1, 5\ s$.*

After exactly one second, that is, at time $t = 1\ s$, we take a second picture. This second picture reveals to us that the cube was, at that precise moment, one centimeter away compared to the previous position, i.e., at the position $x_1 = x_0 + 1 \ cm$ (see Figure 1).

This allows us to conclude that the velocity v of the cube is exactly one centimeter per second: $v = 1 \ cm/s$.

To recapitulate, at time $t = 1 \ s$, we know both the position of the cube and its velocity. In other words, we jointly and simultaneously know the values of these two physical quantities. This enables us to predict with certainty any other position that the cube will occupy in later times.

For example, given that we know that the cube moves with a velocity of one centimeter per second ($v = 1 \ cm/s$), we can predict with certainty that after further 4 seconds, i.e., at time $t = 5 \ s$, it will be exactly in the position $x_2 = x_0 + 5 \ cm$, as is easy to confirm by taking one last picture, just at that moment (see Figure 1).

3. HEISENBERG UNCERTAINTY PRINCIPLE

Let me consider now the famous *uncertainty principle* (which should more properly be called, as it is the case for example in Italian, *indetermination principle*) introduced in 1927 by the German physicist Werner Heisenberg [3]. What does this principle exactly tell us? Well, simply that, contrary to what we have just learned in relation to the wooden cube:

There is no way to determine simultaneously, with arbitrary precision, both the position (x) and the velocity (v) of a microscopic particle, not even by using the most sophisticated measuring instrument!

There is of course no contradiction between this principle and our previous experiment, given that a wooden cube is not a microscopic entity, but rather a macroscopic one, i.e., a body of large dimensions. We can state *Heisenberg uncertainty principle* (HUP) in a bit more precise way with the aid of a very simple mathematical relation. This relation states that the smallest error $Er(x)$ with which we can determine the position x of a microscopic particle, at a given instant, multiplied by the smallest error $Er(v)$ with which we can determine, at the same instant, its velocity v, must always be, approximately, equal to a specific constant c.

In mathematical language, what I have just stated is summarized in the following relation (the symbol "$\cong$" means "approximately equal to"):

$$Er(x) \cdot Er(v) \cong c$$

To give an example, for an *electron*, if we measure the error on the position in centimeters (cm), and the error on the velocity in centimeters per second (cm/s), the value of the constant c is approximately 1 centimeter squared per second (cm/s).

To better understand the content of this relation, we can visualize it graphically, by representing it as a curve, so that only the points that lie on the curve satisfy the HUP (see Figure 2). Let us choose among them the point which is closest to the origin. As you can see, it corresponds to the situation where we have reduced the most, at the same time, both the position error and the velocity error.

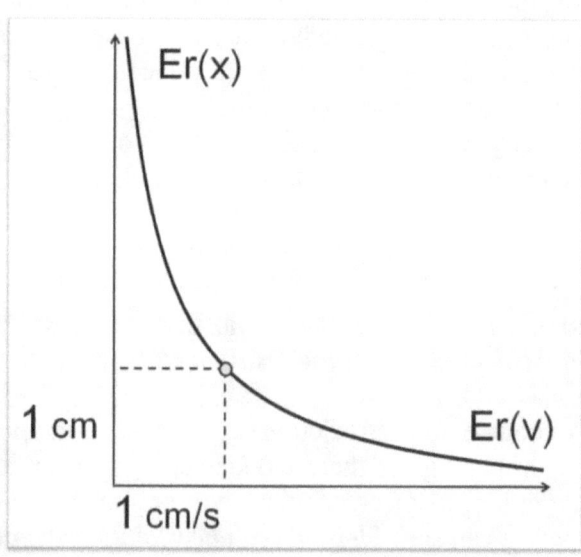

Figure 2. *The points on the curve obey the HUP (here for an electron). The point which is evidenced, with coordinates (1,1), is the one which reduces at best both the error on the position x and on the velocity v.*

But what if we wanted to further reduce, say of one-tenth, the error on the determination of the electron's position?

To do so, and since we are forced to move on the curve, we must evidently slide the point to the right. But in doing so, while reducing the error on the position of one-tenth, at the same time we will increase by a factor of ten the error on the velocity (see Figure 3).

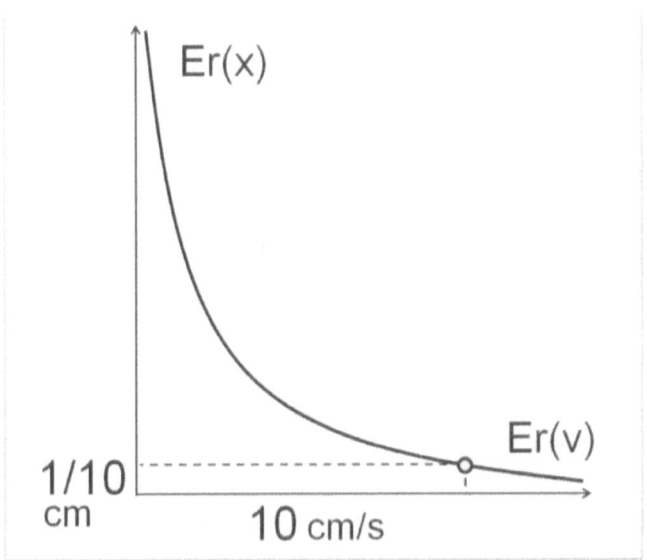

Figure 3. *Reducing by one-tenth (1/10) the error on the position causes the error on the velocity to increase by a factor of ten (10).*

Same thing if instead of the position we try to reduce the error on the determination of the velocity: reducing by one-tenth the error on the velocity will cause the error on the position to increase by a factor of ten.

Very good, but let us now try to understand why Dr. Heisenberg invented, so to speak, his beautiful principle (which is not actually a true principle, as it can be deduced from more fundamental axioms of quantum theory).

Let us first clarify what it means to *see* a macroscopic body.

Consider once again the wooden cube. If we want to see it, we must necessarily light it up with a light source, such as a flashlight. When the light rays strike the cube, they are deflected toward the detecting instrument, which in the present case is your eye, or better your eye-brain system (see Figure 4).

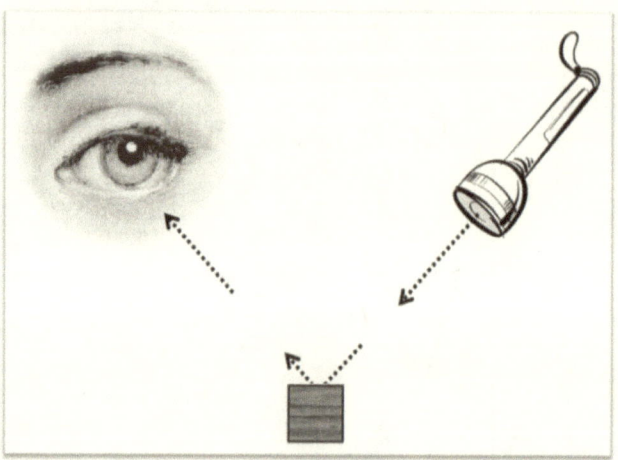

Figure 4. Seeing means observing scattered light.

In other words, to see an object means, roughly speaking, to detect the light coming from that object. We can observe that when the flashlight illuminates the cube, the latter is not in any way disturbed by it. This is therefore a *non-invasive* observational process, through which we are able to *discover* what already existed, regardless of our observation.

Here we can discover not only the existence of the cube itself, but also its characteristics, like its shape and color, and of course its specific location in space. And as with the simple experiment of Section 2, by our observation we are also able to jointly determine the position and the velocity of the cube, without disturbing it.

With microscopic entities, however, this is no longer possible. To understand why, we must first investigate some of the characteristics of light waves.

Light does not exactly behave like infinitely thin and straight rays, but more like waves, and more specifically like *waves of an electromagnetic nature*. Waves are characterizable by some specific pa- rameters. In the case of the so-called *plane waves*, one of these parameters is the wavelength, usually represented by the Greek letter λ (lambda).

The wavelength λ is nothing more than the distance between two successive peaks of the plane wave (see Figure 5).

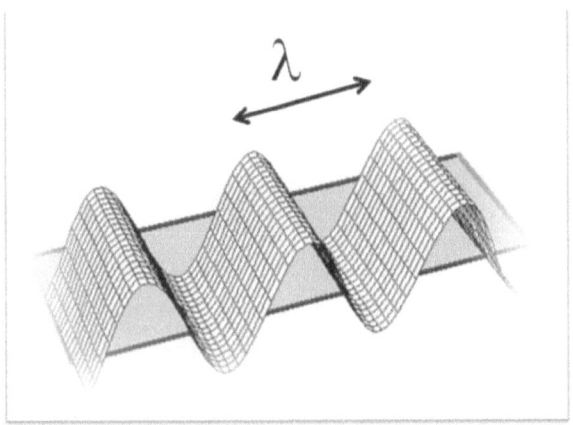

Figure 5. *The wavelength λ of a plane wave is the distance between two of its successive peaks.*

Let us now ask: *What happens when a wave of wavelength λ encounters an obstacle of dimension d?* Well, if the size d of the obstacle is small compared to the wavelength λ, typically nothing will happen, in the sense that the wave will propagate undisturbed, as if the obstacle wouldn't exist.

Let me consider a very simple example: the waves of the sea that pass under a large pier. The poles on which rests the pier are here the obstacles. As you can observe from Figure 6, the waves propagate toward the shore totally oblivious of the poles, in the sense that in no way the poles are able to deflect their direction of propagation.

We are here in the typical situation where the obstacle's size d is small compared to the wavelength λ of the wave ($d \ll \lambda$), so that the latter cannot be detected by the former.

Figure 6. *When the size d of the obstacle is small with respect to the wavelength λ of the wave, the latter experiences almost no deviations in its direction of propagation.*

What happens instead if the size d of the obstacle is large compared to the wavelength $λ$ of the wave? To answer, consider the example of a small island.

As you can see on the drawing of Figure 7, which offers a top-down perspective, the wave coming from the north, winds along the two sides of the island, thus changing its direction of propagation. In this way, behind the island, a "shadow zone" results, where the wave interferes with itself.

We are here in the typical situation in which the obstacle's size d is large compared to the wavelength $λ$ of the wave ($d \gg λ$), and because of that is able to modify in a detectable way its motion.

Thanks to the above two examples, it should be intuitively clear now that the wavelength $λ$ of the wave used to see an object poses a clear limit to the precision with which it will be possible to locate it in space.

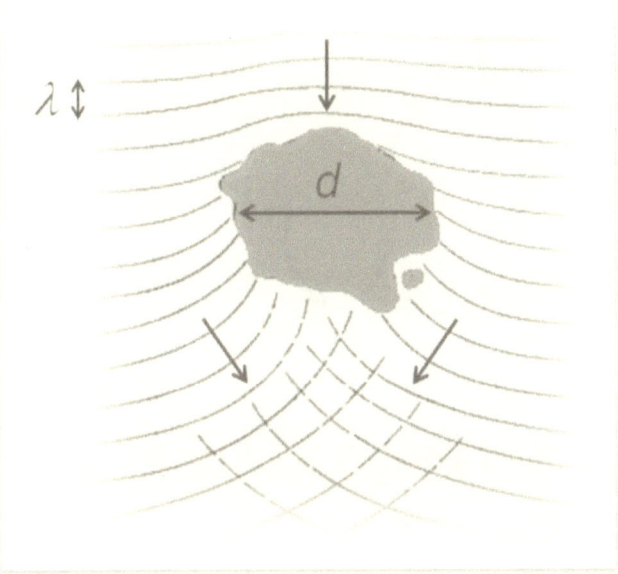

Figure 7. *When the size d of the obstacle is large compared to the wavelength λ of the wave, the latter will experience important deviations in its propagation direction.*

In fact, if λ is too large compared to the size d of the object, the wave will not be deflected by the same, and we will have no way of knowing about its presence. This means that:

The resolving power of an optical instrument can never be greater than the wavelength of the radiation used to illuminate the different objects.

The *resolution* of an optical instrument, however, depends not only on the wavelength, but also on the *angular aperture α* of the instrument (see Figure 8), because of so-called and well-known *refractive phenomena*, which are able to blur the image of objects, thus placing a limit to the details that can be distinguished. In other words:

The greater is the angular aperture α of an instrument and the better will be its resolution.

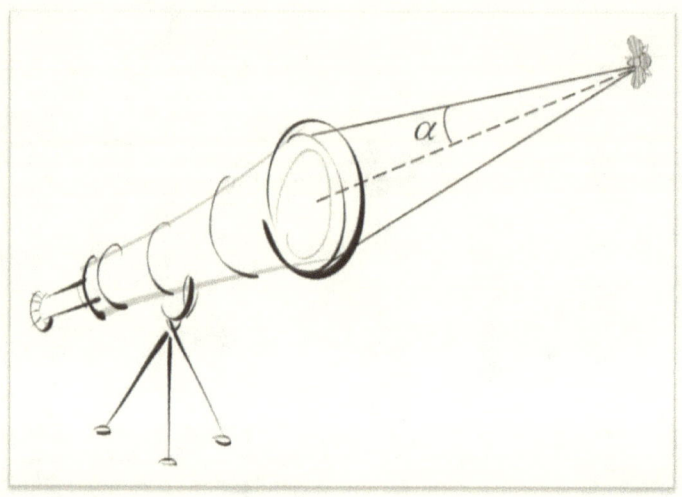

Figure 8. *The opening angle α of an optical instrument.*

So, summing up, the resolving power of an optical instrument, such as a microscope, depends both on the wavelength λ used and on the angular aperture α of the lens.

We can synthesize all this with the following simple mathematical expression, stating that the minimum error $Er(x)$ we commit in determining the position x of a body, because of the limited resolving power of an instrument, is directly proportional to the wavelength λ of the radiation used, and inversely proportional to the *sine*[2] of its angular aperture α:

$$Er(x) = \frac{\lambda}{sen\ \alpha}$$

As you can see from the above simple relation, if you want to reduce the error in the determination of the position x, a possible strategy is evidently that of reducing the wavelength λ of the

---

[2] If you have never heard of the *sine* function in trigonometry, do not worry, it does not really matter for the continuation of our reasoning.

radiation used.

This is of course always possible, as an entire *electromagnetic spectrum* is available, virtually infinite, ranging from *radio waves* of long wavelengths up to so-called *gamma rays*, whose wavelengths are very small (see Figure 9).

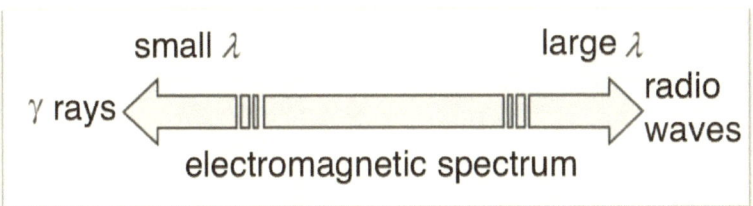

Figure 9. *The electromagnetic spectrum consists of the set of all the electromagnetic radiations of different wavelengths λ.*

So, using a short wavelength radiation of the gamma (γ) type, it should then be possible, at least in principle, to detect the position (x) of the tiniest corpuscles, such as the electrons.

Now, as we have seen, to see where an object is, you have to irradiate the object with an electromagnetic wave, then look at the scattered wave. In other words, the wave has to interact with the object. *But what does it mean, in this specific context, to interact?*

To fix ideas, let us first consider the simple case of two marbles, able to slide on a plane without friction and without rotating. The first, of black color, is immobile, while the second, of white color, impinges the first with velocity v (see Figure 10).

As you can observe in Figure 11, following the interaction, i.e., following the collision, the white marble has exchanged a certain amount of *momentum*, and consequently a certain amount of *energy*, with the black marble, which therefore has been set in motion, with a certain *scattering angle*.

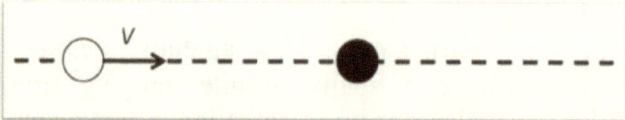

Figure 10. *The situation before the collision: the white marble moves towards the black one, with speed v.*

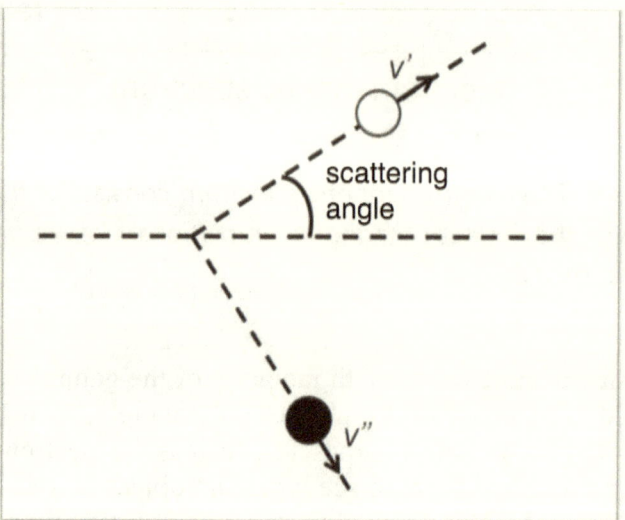

Figure 11. *The situation following the collision: the white marble moves with speed v', lower than the initial speed v, and a given scattering angle; the black marble, which received a certain amount of momentum, also moves with some speed v''.*

So far, everything seems clear, but what happens if the black marble, instead of being a macroscopic body, is an elementary particle, such as an electron (which for convenience I will still represent like a marble), and the white marble is not a corpuscle, but an electromagnetic wave (see Figure 12)?

In this case, the situation after the collision is like the one represented in Figure 13.

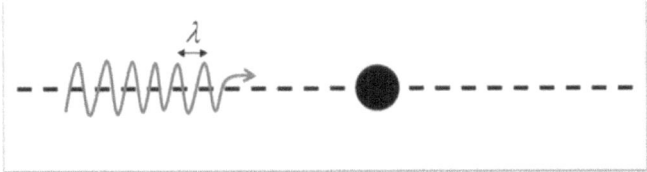

Figure 12. *The situation before the collision: the electromagnetic wave (of wavelength λ) propagates (at the speed of light) in the direction of the electron.*

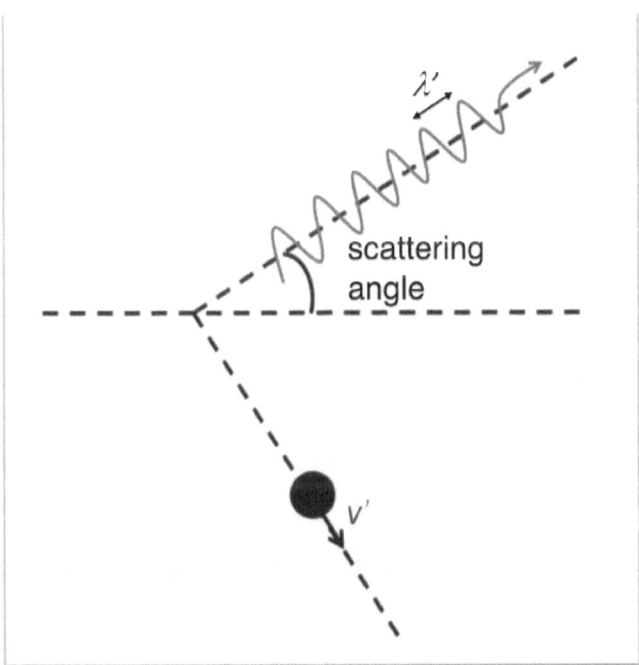

Figure 13. *The situation after the collision: the electromagnetic wave is scattered with a given scattering angle and with a wavelength λ' that is longer than the initial wavelength λ.*

If you compare Figure 13 with Figure 11, you will notice that the interaction process looks a lot like the previous one: the incoming wave, as if it were a marble, communicates to the electron a certain amount of momentum, also in this case by setting it in motion.

But look more closely at what happens to the scattered wave (Figure 13): the wavelength λ of the incident wave, following the interaction, has changed, in the sense that the wavelegth λ' of the scattered wave is longer than λ ($\lambda' > \lambda$).

This effect of increase of the wavelength is called the *Compton effect* (or *Compton shift*), because it was discovered by the American physicist *Arthur Compton* in 1923 [4].

What I want here to highlight is that electromagnetic waves, similarly to moving marbles, also possess a certain amount of momentum, and when they interact with elementary particles, such as electrons, they can transfer to them part of their momentum. And when this happens, their wavelength change, in the sense that it increases.

This means, among other things, that waves possess a momentum p which is inversely proportional to their wavelength λ. Translating this observation in mathematical terms, we can write:

$$p = \frac{h}{\lambda}$$

where the constant of proportionality h is the famous *Planck constant*, whose value is really very small (approximately $6{,}626 \cdot 10^{-34} \, J \cdot s$).

To exactly determine the value of the momentum transferred to the corpuscle, it is not sufficient, however, to just know the wavelength of the scattered wave. You also have to know the scattering angle.

This is because momentum, like velocity, is a vector quantity, and a *vector* quantity can vary for two distinct reasons: because its *numerical value* varies, or because its *direction* varies.

If you do not fully understand this, no problem, it is not an essential point for understanding what follows. But what you have to keep in mind is that:

To determine the momentum transferred to the corpuscle, you must also determine the scattering angle.

Now, the precision with which you can determine the scattering angle is limited by the angular aperture of the lens of the optical instrument you use (see Figure 8).

With a simple geometry reasoning (which I leave it to the reader with some basic knowledge of trigonometric functions), it is easy to see that it is not possible to determine the amount of momentum transferred to the microscopic particle with an error $Er(p)$ lesser than the momentum p of the incident wave multiplied by the sine of the diffusion angle α. More precisely:

$$Er(p) \cong p \cdot sen\, \alpha$$

As we have just seen, the momentum of a plane wave is simply given by the Planck constant h divided by the wavelength λ. If you use this in the above expression, you obtain:

$$Er(p) \cong \frac{h}{\lambda} \cdot sen\, \alpha$$

At this point, if you remember that the momentum of the particle is given by the product of its mass m times its velocity v ($p = m \cdot v$), you can divide by m on the right and left sides of the above expression, thereby obtaining an estimate of the minimum error $Er(v)$ on the velocity:

$$Er(v) = \frac{Er(p)}{m} \cong \frac{h}{\lambda \cdot m} \cdot sen\, \alpha$$

On the other hand, considering the previously derived expression for the minimum error $Er(x)$ on the position of the particle, multiplying it by the obtained expression for $Er(v)$, you get:

$$Er(v) \cdot Er(x) \cong \frac{h}{\lambda \cdot m} \cdot sen\, \alpha \frac{\lambda}{sen\, \alpha} = \frac{h}{m}$$

This is nothing but the previously stated state *Heisenberg uncertainty principle* (HUP), with the constant $c = h/m$ (not to be confused with the speed of light) whose value for an electron is about $1\ cm^2/s$. In other words, you have just derived the famous HUP.

4. COMPLEMENTARITY AND INCOMPATIBILITY

The strange situation expressed by the HUP (and many other situations that are encountered when one studies microscopic systems) has been summarized in 1928 by the Danish physicist *Niels Bohr* in his famous *principle of complementarity* [5]. Roughly speaking, this principle states that:

There are properties that are mutually exclusive, and therefore cannot be observed simultaneously, using a same experimental arrangement, i.e., within the same experimental context.

So, if we measure (that is, if we observe, in a practical way) with good precision the position property of a particle, we will automatically and inevitably alter in a profound and totally unpredictable way (or even make it indeterminate) its velocity property, and vice versa. The two properties – possessing a given position and possessing a given speed – being in a sense complementary, they cannot be jointly observed.

At this point, two remarks are in order. The first is that this possible alteration, for example of the velocity when we observe the position, takes place in a completely unpredictable way, that is, in a way that is not determinable a priori by the observer. This aspect of the unpredictability does not emerge directly from our simplified analysis of the Compton Effect. However, it is an integral part of the formalism of quantum mechanics.

To put it simply, according to quantum theory:

We cannot determine in advance what will be, for example, the scattering angle following the collision, and can only calculate the probabilities associated with the different possible scattering angles.

The second remark is that in all our reasonings we have implicitly assumed that the microscopic particle always possessed, even before being observed, a specific position and a specific velocity, although these were not known by the observer/experimenter.

As we are going to see, this assumption is however completely unfounded.

Having said that, let us try to understand the concept of complementarity a little better. The word "complementarity" is obviously quite attractive from a philosophical point of view, and in part is certainly correct, but it can also lead to a possible misrepresentation of the issue we are analyzing here.

Instead of the term "complementarity," you can use the simpler and more direct term of "incompatibility," to be understood in the sense of the incompatibility of the procedures of observation of certain properties, associated with specific experimental arrangements.

In physics, one tries to make precise the concept of experimental incompatibility by using the idea of *non-commutability*. More precisely:

If two observations are compatible, the order with which you perform them does not affect their outcomes (and therefore such order can be freely switched). When instead a change in the order of the observations can affect their outcomes, the two observations are said to be incompatible.

This is exactly what happens with the position and the velocity of a microscopic particle: to observe first the position and then the velocity does not produce the same results than to observe (i.e., to measure) first the velocity and then the position. This is mainly due to the fact that these observations are invasive processes, modifying the state of the observed entity in an unpredictable way.

It is however important to understand that the incompatibility I'm here talking about is not a feature of the microscopic processes only: it can also manifest in many of the operations we perform every day. Let me consider a simple example.

I hope you will agree that to put on the socks first, then the shoes, does not produce the same outcome as to put on the shoes first, then socks (see Figure 14).

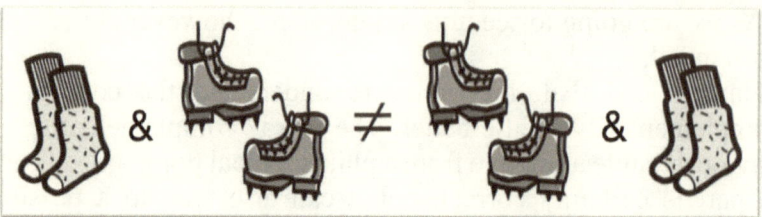

Figure 14. *The operations "putting on the socks" and "putting on the shoes" are non-commutative (the symbol "≠" means "not equal").*

These two processes being *non-commutative* (their order of execution is crucial for the final result), they can be considered to be mutually incompatible. But of course, not all processes are mutually incompatible. Many are perfectly compatible. Let me consider a simple example of two perfectly compatible operations, that is, two operations whose order of execution can be switched, without affecting the final result.

I hope you will agree that to put on the socks first then the gloves, is the same, i.e., it produces the same result, than to put on the gloves first then the socks (see Figure 15).

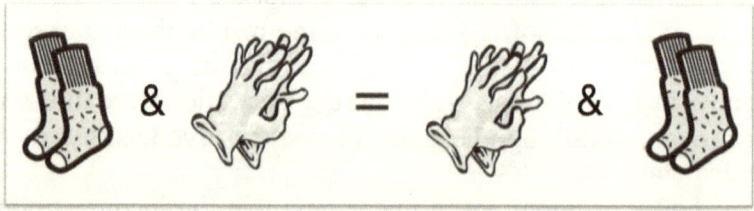

Figure 15. *The operations "putting on the socks" and "putting on the gloves" are commutative.*

It is instructive to consider some further examples of operations that are mutually incompatibles.

A novice asked the prior: "Father, can I smoke when I pray?" And he was severely reprimanded. A second novice asked the prior: "Father, can I pray when I smoke?" And he was praised for his devotion.

In other words, to pray and smoke does not produce the same effect as to smoke and pray (see Figure 16).

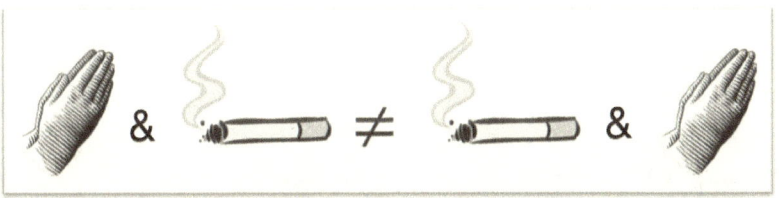

Figure 16. *The operations "to pray" and "to smoke" are non-commutative, according to the prior's understanding.*

Here the non-commutability is expressed through the order chosen for the verbs "to smoke" and "to pray" in a sentence.

If you switch the order of the verbs, it also changes the perceived sense of the phrase. This because verbs indicate actions, that is, operations that we perform, which is the reason why their order in a sentence is often so crucial.

What we must understand is that, in general, the order with which we operate in reality affects the final outcome. To assemble an IKEA piece of furniture, it is necessary to operate in exactly the sequence indicated in the instruction manual, if you want to obtain the desired result.

To make sure that this issue is fully understood, let me consider still another example, using a simple *right triangle* (a triangle in which one angle is 90°). I define the operation A as consisting in rotating the triangle 90° clockwise. The operation B, instead, is by definition a reflection of the figure with respect to the vertical axis. As you can see in Figure 17, depending on the order of the operations, the final result will not be the same: A and B are therefore incompatible operations, being non-commutative operations.

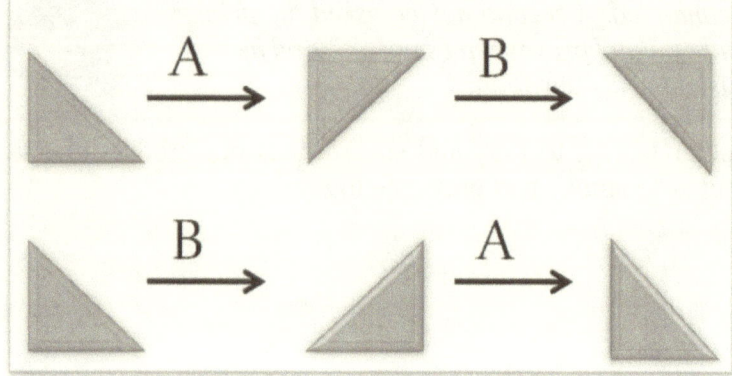

Figure 17. *The "rotation" and "reflection" operations are non-commutative, as depending on their order they will in general not produce the same result.*

5. MEASURING A WOODEN CUBE

Having clarified the concept of incompatibility, and the fact that incompatibility can be expressed in terms of non-commutability, let me show what are the consequences of all this when you try to observe two specific properties of an ordinary macroscopic entity, to you now familiar: a wooden cube [6].

So, the physical system you are about to study/observe is a simple wooden cube. With the letter A, you decide to denote the process of observation of the property of the cube of *burning well*. On the other hand, with the letter B, you also denote the process of observation of the property of the cube of *floating on water*.

There are of course different possible ways to define these two properties of burning well and floating on water. So, if we want to be more precise, we must specify what you mean in practice, for the cube, to have these two properties tested, i.e., what are the operations you have to exactly perform, and the results you have to obtain, to successfully observe these two properties.

For example, you can decide that the observation of the burning well property involves exposing the cube to the flame of a match, for a few seconds. If, following this operation, the cube is set on fire and reduces to ashes, the observation of the burning well property is considered to be successful, and you can say that the property has been *confirmed* (see Figure 18).

You can also decide that the observation of the floating on water property consists in completely immersing the cube in a container filled with water and then check if thanks to Archimedes' *buoyant force* it raises to the surface. If this happens, the observation of the floating property is considered to be successful, and again you can say that the property has been confirmed (see Figure 18).

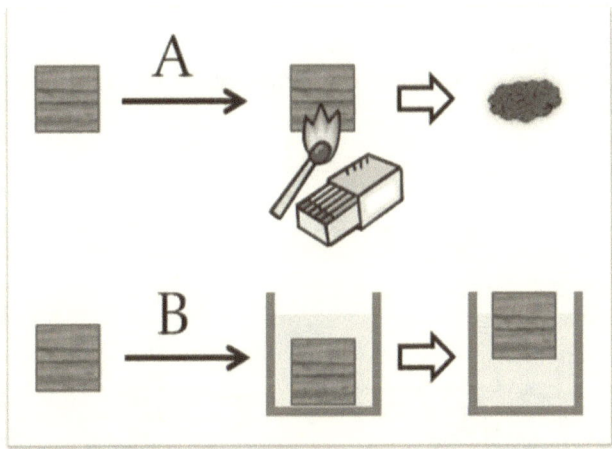

Figure 18. *Operations A and B, and their respective outcomes, when carried out on a wooden cube.*

Considering our previous discussion, it is then natural to ask the following question:

Does the wooden cube possess <u>both</u> properties: burning well <u>and</u> floating on water? (Which for reasons of conciseness, we will also call burnability and floatability, respectively).

One way to possibly verify this consists in taking the wooden cube and then try to observe these two properties, one after the other. If you start with burnability, you can see that the cube burns well, i.e., that it turns into a small pile of ashes. But then, if you try to observe its floatability, plunging the obtained ashes into the water, these will not raise to the surface, since, as is well known, ashes do not float (see Figure 19).

So, the conclusion of the above sequence of observations is that the wooden cube burns well, but does not float.

You could then try reversing the order of the two observations. If you start with the floatability, you can see that the wooden cube floats easily. However, if you subsequently want to observe its burnability property, subjecting it to the flame of the match, it will not burn, since a wet cube, as is known, does not burn well (see Figure 19).

So, the conclusion of this second sequence of observations is that the wooden cube floats on water, but does not burn well.

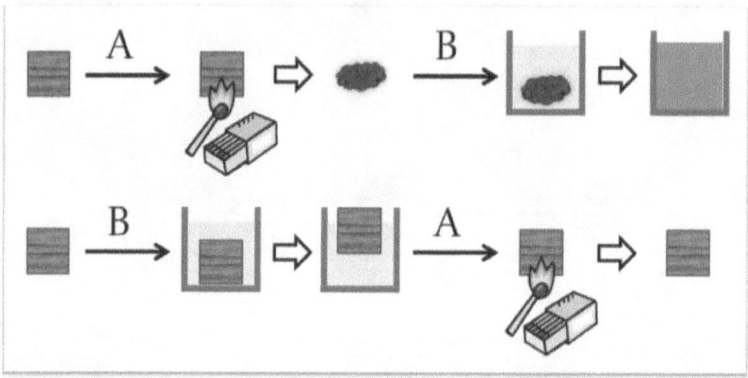

Figure 19. *Operations A and B, depending on their order of execution, produce different outcomes. If A is successful, then the outcome of B will be negative, and vice versa, if B is successful, then the outcome of A will be negative.*

What you have just pointed out is the simple fact that the observational processes *A* and *B*, associated with the burnability and floatability properties, respectively, do not commute. In other words, they are mutually incompatible. Therefore, it does not seem possible to jointly observe the burnability and the floatability of the wooden cube. If this is true, as it is true, then you may wonder:

Does the wooden cube really possess, at once, both properties of burnability and floatability?

The question is legitimate, since apparently you are unable to jointly observe these two different properties. On the other hand, according to your intuition, the wooden cube certainly possesses at once both properties of burning well and floating on water. In the same way for example a car can be at the same time crash-proof and 4 meters long. More precisely:

Intuition tells you that an entity can possess at once a number of different properties, al- though not all of them are necessarily observable at the same time, or one after the other.

Very well, let us recap. It is clear that the cube possesses the property of burning well. It possesses such property because if you execute the *test A* that by definition allows to observe it, the test will invariably be successful, so the property will be confirmed.

In the same way, it is clear that the cube possesses the property of floating on water. It possesses such property because if you execute the *test B* that by definition allows to observe it, the test will invariably be successful, so the property will be confirmed.

Perfect, but since your intuition also tells you that the cube possesses the *meet property* of burning well and floating on water, it is natural to ask:

What would be the test C allowing to confirm the meet property of burning well and floating on water? In other words: How can you know if your 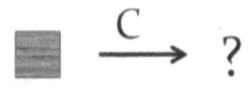 *intuition is correct and that it is true that the cube possesses at once these two properties, despite the fact that they are mutually incompatible?*

Apparently, you are confronted here with a little puzzle. Indeed, if you take a look at the *truth tables of classical logic*, and more particularly the truth table for the *conjunction*, described in Figure 20, you can observe the following.

Reading the first line of the table, you find that if a property *A* is false, and a property *B* is false, then, inevitably, also the meet property "*A and B*" is false. In other words, in classical logic the meeting of two falsities is once again a falsity.

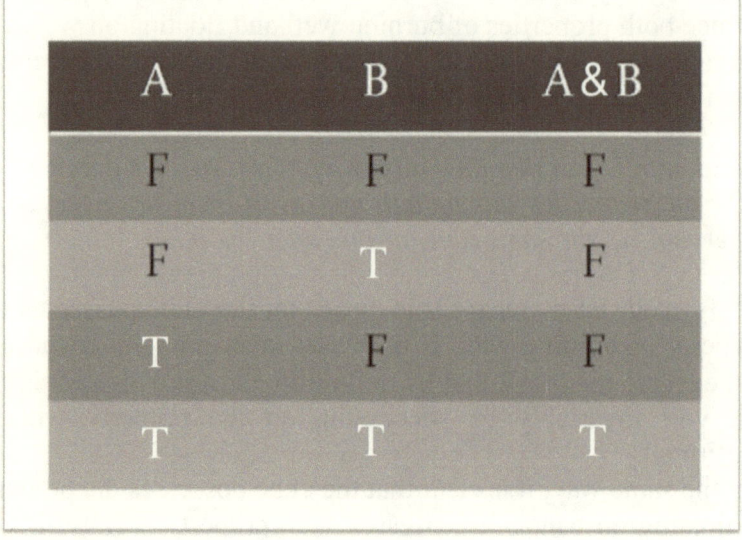

Figure 20. *The truth table of classical logic, here for the conjunction logical operator "and."*

But, as evidenced by the second and third line of the table, it is also sufficient that only one of the two properties is false, to make the associated meet property, as a whole, false.

In fact, as made evident in the last line of the table, the meet property "*A and B*" can be true *if and only if* both property *A* and *B* are individually true.

On this obviously you can only agree. But considering that burnability and floatability are mutually incompatible properties, how can you bring out their joint trueness. In other words:

How can you test together, conjunctly, at once, the truth value of two properties that are experimentally mutually incompatibles?

I hope it is clear what is the relevance of this discussion in relation to the previous analysis of Heisenberg uncertainty principle (HUP), on which of course I will be back shortly. In fact, if you remember, the position and velocity of a microscopic particle are linked by an uncertainty relation; a relation which in turn expresses a condition of incompatibility.

Does this mean that a microscopic particle is not able to jointly possess a position and a velocity? In other words:

Is HUP a statement about the simultaneous non-existence of the position and velocity of a microscopic entity, such as an electron, or is it just a statement about our limitation in jointly knowing these two physical quantities?

Based on your intuition about the burnability and floatability of a wooden cube, I'm sure you would be tempted to say that the experimental incompatibility of two physical quantities does not mean that they cannot exist simultaneously, and that therefore nothing prohibits an electron to simultaneously possess a well-defined position and velocity.

Is the above correct? To find out, I propose to continue our conceptual analysis, and for this it will be useful to define a bit more clearly what a *property* is.

6. THE EPR-PA REALITY CRITERION

In general, you can say that a property is something that an entity can have, which can be observed, and is defined, by means of an *experimental test*, whose execution enable one to confirm (or to invalidate) the property in question. But be careful:

To confirm a property does not necessarily mean to prove that the property is or was actual, in the sense that the property is or was stably possessed by the entity in question.

As the wise would say: *a burnt wooden cube is no longer a burnable wooden cube*!

That said, to continue your exploration, you now need the help of the German physicist *Albert Einstein* and of his two Russian and American-Israeli collaborators, *Boris Podolsky* and *Nathan Rosen* (see Figure 21).

In a famous article published in 1935 [7], these three scientists enunciated an important *reality criterion*, which is of course also an *existence criterion*, since in the common understanding of these two concepts, something is considered to be real if and only if it exists.

More precisely, Einstein, Podolsky and Rosen (in brief, EPR), in their famous article, said the following:

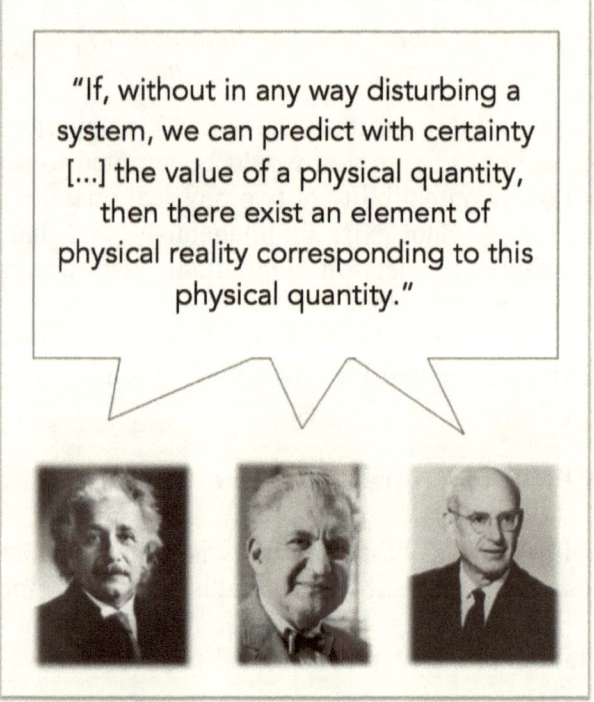

"If, without in any way disturbing a system, we can predict with certainty [...] the value of a physical quantity, then there exist an element of physical reality corresponding to this physical quantity."

Figure 21. *From left to right, physicists Albert Einstein, Boris Podolsky and Nathan Rosen.*

What EPR have clearly recognized is that our description of re-ality is essentially based on our reliable predictions about it. However, they also remained rather cautious regarding their crite-rion, as they added in their article that:

"It seems to us that this criterion, while far from exhausting all possible ways of recognizing a physical reality, at least provides us with one such way, whenever the conditions set down in it occur. Regarded not as a necessary, but merely as a sufficient, condition for reality, this criterion is in agreement with classical as well as quantum-mechanical ideas of reality."

Despite their warning, EPR did not offer a single counter example of what would be the nature of an element of physical reality not subject to their criterion. In other words, although they assumed, very prudently, that their criterion was only sufficient, they presented no reasons as to why it should not be considered, at least in principle, also necessary.

But let me come back once more to the wordings of the criterion. An important point to emphasize is that when EPR write "if [...] we can predict with certainty," what one should understand is: "if we can in principle predict with certainty."

Indeed, the important point is not if one possesses in practice all the information allowing to make a reliable prediction, but if this information is available somewhere in the universe (although maybe dispersed who know where), so that a being of sufficient power and intelligence could in principle access it.

That said, it is worth observing that this important criterion of reality, or of existence, was subsequently reconsidered by the Belgian physicists *Constantin Piron* [8] and *Diederik Aerts* [6] (see Figure 22), the latter being also the author of the paradigmatic example of the floatability and burnability of a wooden cube.

These two physicists reformulated the EPR criterion in a much more specific and complete form, which is roughly the following [6, 9, 10]:

> If, without in any way disturbing the physical entity under consideration, it is in principle possible to predict with certainty the successful outcome of an experimental test, then the property associated with that test is an actual (existing) property of the physical entity. And vice versa.

Figure 22. From left to right, physicists Constantin Piron and Diederik Aerts.

In other words, if the property of a physical entity is actual, then (without in any way disturbing the entity) it is in principle possible to predict with certainty the successful outcome of an observational test associated with it. Therefore, according to this more complete reality criterion, which I will simply call the *Einstein-Podolsky-Rosen-Piron-Aerts* (EPR-PA) *criterion*: a property is actual if and only if, should one decide to perform the observational test that defines it, the expected result would be certain in advance.

This means that the entity has the property in question before the test is done, and in fact even before one would have chosen to execute it, which is the reason why one is allowed to say that the property is an *element of reality*, existing independently from our observation.

On the other hand, if one cannot apply the EPR-PA criterion, i.e., if one cannot, not even in principle, predict the outcome of the test defining the property in question, one must conclude that the entity under consideration does not possess that property, i.e., that the property is not an actual (existing) one, but only a *potential* property (if the probability of actualizing the property, in some experimental contexts, is non-zero).

Again, the above conclusion is correct provided the prediction cannot be made even in principle. Indeed, in most experimental situations one simply does not possess a complete knowledge of the entity, and therefore one does not have access to all its actual properties.

But when one possesses a complete knowledge of the entity, then by definition one is also able to predict with certainty all that is predictable about it, so that what cannot be predicted is, by definition, a non-existing (potential, uncreated) aspect of reality.

After this important detour on the issue of reality criteria, it is time to return to our little wooden cube. Based on the EPR-PA criterion, to determine whether the wooden cube possesses or not the property of burning well, you do not have to execute the burnability test, and burn it, but simply be in a position to predict with certainty that, should you perform the test, the outcome would be certainly positive.

Consider a much more straightforward example: think about the city of *Venice* in *Italy* (see Figure 23), now, and ask yourself:

Does Venice exist right now?

I'm assuming that in this moment you are not in Venice, that is, that you are not having an experience with Venice in this moment. In other words, you cannot base your claim about the existence of Venice now, on the fact that you would be having, now, an experience with it.

But this is not necessary, as per the EPR-PA criterion, to be able to affirm that Venice exists in this moment, all you need is to be able to predict that, should you have decided to have an experience with Venice now (for example by organizing a trip a few days ago, to go to Venice, so that you could be there now), the latter, with certainty, would have been available to be part of it.

Figure 23. *A typical postcard image of Venice (Italy).*

So, summing up:

Reality (what exists) does not only correspond to the phenomena that you factually experience. Reality also and above all corresponds to all the possible phenomena: those that you could have in principle experienced with certainty, if only you would have chosen to do so in your past.

Reality, therefore, is constructed in a *counterfactual* way. In the sense that you can speak in a perfectly meaningful way even of things you are not currently concretely observing, provided that, if you would have decided to do so, the positive result of the observational process would have been absolutely certain in advance.

How can you apply all this to the problem of our wooden cube? More particularly:

How can you use this to solve the problem of demonstrating that the wooden cube actually possesses the meet property of burning well and floating on water, although these two properties are individually experimentally mutually incompatible?

Well, according to the EPR-PA criterion, it is sufficient to be able to predict that, should you perform the test C associated with such meet property, the positive outcome would be certain in advance.

All right, but:

What would be then this mysterious test C, associated with the meet property of burning well and floating on water? In other words: What is, generally speaking, the observational test of a meet property "A and B"?

The specifications of this particular test, which in technical language is called a *product test*, were given some years ago by Constantin Piron [8, 9], who I mentioned in relation to EPR's reality criterion. Let me explain what a product test is.

It is very simple: first of all, you need an instrument that can generate, in a completely *random* way, two events, which I will simply call "heads" and "tails." For example, the instrument could be the toss of a coin, provided it is carried out in such a way as not to allow you (the experimenter) to predict the outcome in any way.

So, if following the toss of the coin you obtain "heads," you perform *test A*, and the outcome (positive or negative) will be assigned to the *test C*, of the meet property. On the other hand, if you obtain "tails," you perform *test B*, and again the outcome (positive or negative), will be attributed to C (see Figure 24).

Here you are, now you know the specific nature and logic of the observational test of a *meet property*, called a *product test*.

Perhaps you will object that, since following the coin toss you only have to perform one of the two tests, this means that only one of the two properties would in fact be tested.

This is not exact, as you do not have to forget that the toss of the coin is an integral part of the observational process, and that the choice of which of the two tests will be performed is *totally unpredictable*.

Therefore, the only way to ensure *a priori, with certainty*, the positive outcome of test *C* (without the need of executing it), is to have the cube possessing both properties at once, that is, possessing them at the same time!

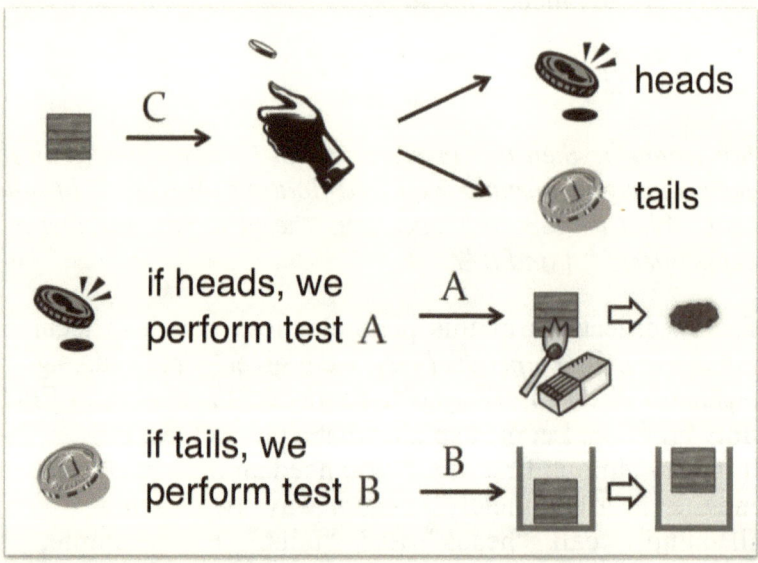

Figure 24. *Graphical representation of the execution logic of a product test, able to test the meet property of "burning well and floating on water."*

So, when you go from properties to tests, the conjunction logical operator "and" transforms into the disjunction logical operator "or:" to test the meet property *A and B*, you have to test either *A or B*, choosing however one of these two alternatives – and this is really the crucial point – totally at random.

That said, what can you say then about the meet property of the cube of jointly burning well and floating on water? Does it possess it, in actual terms, or not? Evidently, according to the test product C just defined, and to EPR-PA reality criterion, as you are able to predict with certainty the positive outcome of both tests, of burnability and floatability, you can deduce that the wooden cube jointly possesses these two properties, at once, although they are mutually experimentally incompatible.

In summary:

The experimental incompatibility of two properties A and B of a given entity does not necessarily imply that they cannot be simultaneously actual, as the disjunctive logic of a product test shows, and the example of the wooden cube demonstrates.

As a result, you may now be tempted to conclude that, even though the position and velocity of a microscopic corpuscle are incompatible physical quantities, as expressed by HUP, nevertheless, it should be possible to consider them to be simultaneously actual.

But such consideration would be completely wrong!

7. NON-SPATIALITY

Let me try to explain why the position and velocity of a microscopic particle, contrary to the burnability and floatability of a wooden cube, cannot be considered properties that are jointly possessed by a microscopic entity like an electron.

For this, let me start considering again the case of a macroscopic body, that is, of a body of large dimensions, visible with the naked eye. At time t_0, the body is, say, located in a position x_0. and, at that same instant, it also possesses a well-defined velocity v_0 (see Figure 25).

As previously discussed, you know that when you both know the position (x_0) and the velocity (v_0) of a body, at a given instant (t_0), you can then calculate (thus predict with certainty) any other position and velocity that the body will occupy, in any subsequent time, by solving the so-called *equations of motion*. You can for instance determine its position x_1 and velocity v_1, at a subsequent time t_1, or its position x_2 and velocity v_2 at a further time t_2, and so on. This means that the body goes along a *trajectory in space*, as time passes by, which is perfectly defined, i.e., a priori knowable with certainty.

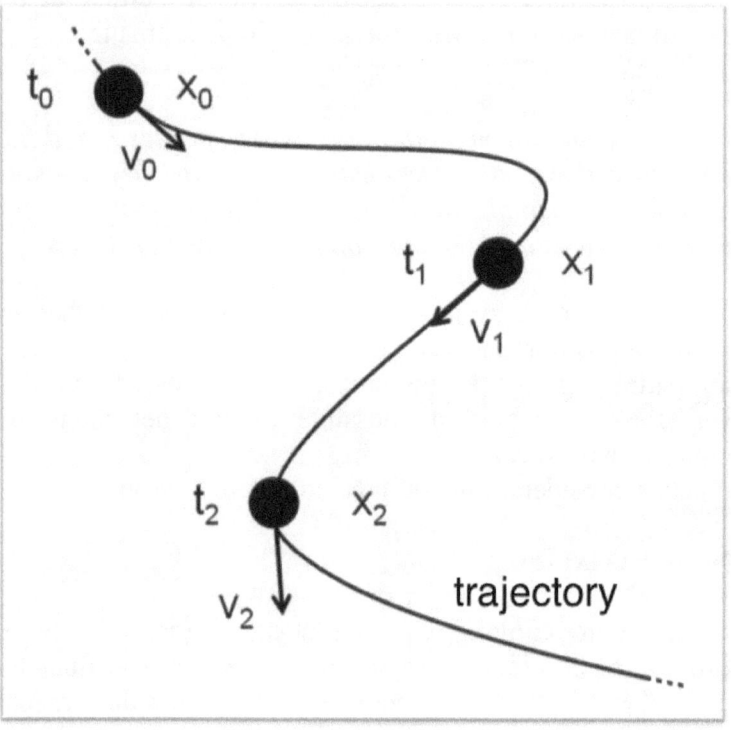

Figure 25. *A graphical representation of the trajectory traveled in space by a moving macroscopic body, i.e., of the different positions taken by the body over time.*

In other words, to solve the equations of motion is equivalent to predict with certainty any future position and velocity of the macroscopic body in question. I will not enter here into the details of these equations of motion, which depending on the physical systems can become quite complex. What is important to understand is that these equations are like a "mechanical device," and when you feed such device with a precise input, formed by a position and a velocity, evaluated at a same instant of time, say at time $t = 0$, the device will invariably provide you with outputs, corresponding to the positions and velocities at any other instant of time t, both in the future and in the past (see Figure 26).

$$\text{input} \quad x_0 \text{ e } v_0 \xrightarrow{\quad} \boxed{\frac{d}{dt}\frac{\partial L}{\partial \dot{q}^i} - \frac{\partial L}{\partial q^i} = 0} \xrightarrow{\text{output}} x_t \text{ e } v_t$$

input
x_0 e v_0 (time 0)

$$\frac{d}{dt}\frac{\partial L}{\partial \dot{q}^i} - \frac{\partial L}{\partial q^i} = 0$$

equations of motion

output
x_t e v_t (time t)

Figure 26. *The equations of motion allow to predict every position and velocity of a macroscopic body, based on a precise input (the so-called initial condition).*

It is by considering this remarkable property of the equations of motion that the Frenchman Pierre- Simon de Laplace (see Figure 31), towards the end of the eighteenth century, enunciated his famous *principle of determinism*, more or less in these words [11]:

If, at a certain moment, we would simultaneously know the position and the velocity of all bodies of the universe, then, in principle, we could predict their behavior at any other time, both in the past and in the future.

For Laplace, the simultaneous knowledge of the position and velocity of all bodies in the universe was entirely possible, at least in principle. However, due to HUP, you know today that he was wrong, that a knowledge of this kind is absolutely unthinkable, and this not for a lack of information, or of an adequate technology.

Indeed, if you remember, HUP does not allow one to jointly determine, with arbitrary precision, the position and velocity of a micro-entity, like an electron, let alone all those in the universe!

So, there is no way to insert in the equations of motion the input required and, consequently, the equations of motion are no longer able to provide you with the desired outputs. Accordingly, you are no longer able to predict, in no way, the position and velocity of the micro-entity under consideration, at any other instant of time t. And this means that the principle of determinism, as it was enunciated by Laplace, is not valid for a microscopic entity.

Failing the ability to determine, i.e., to predict with certainty, the future positions and velocities of a microscopic entity, what can you conclude on the basis of EPR-PA reality criterion?

Well, it is very simple. The criterion tells you that the possibility to predict with certainty the position and/or velocity of a particle is equivalent to the reality, that is, to the existence, of the position and/or velocity of that corpuscle.

But if the possibility of predicting these quantities fails, in the sense that they cannot be predicted, *not even in principle*, this means that they cannot be considered to be real, existing quantities. So, you are forced to conclude the following:

 *Microscopic particles (electrons, protons, neutrons, etc.) do not exist! In the sense that 'they do not exist as particles', i.e., as entities that would be stably localized in space, thus possessing at any moment a well-defined position, velocity and energy!*

If I also say "energy" it is because, as is known, the energy of a material body is in general a function of both its velocity and position. And if these quantities are not actually existing, then the same must be true also for the energy. In short, the so-called "microscopic particles," which particles are not, are non-spatial entities!

In other words, if a macroscopic body is able to possess, in every moment, a well-defined position, velocity and energy, a microscopic pseudo-corpuscle, instead, cannot possess in general such attributes. To say it with the thought-provoking words of *Diederik Aerts*, we must surrender to the evidence that [12]:

> Reality is not contained within space. Space is a momentaneous crystallization of a theatre for reality where the motions and interactions of the macroscopic material and energetic entities take place. But other entities – like quantum entities for example – "take place" outside space, or – and this would be another way of saying the same thing – within a space that is not the three-dimensional Euclidean space.

This means that the three-dimensional space in which we live, with our physical body, macroscopic in nature, is only a small theater, which cannot contain all of our physical reality.

Dimensionally speaking, reality is much bigger than that, and cannot be represented on such a small three-dimensional stage. So, there must be other "stages" out there, able to accommodate entities having a genuine non-spatial nature; entities whose spatiality is of a very different, non-ordinary kind.

But if the microscopic entities generally do not have a position, what does this exactly mean? How can one understand the process through which a physicist, under certain experimental conditions, can observe the spatial position of an elementary entity? The answer given by Diederik Aerts is simple [12]:

"Quantum entities are not permanently present in space [...] when a quantum entity is detected in such a non-spatial state, it is 'dragged' or 'sucked up' into space by the detection system."

Therefore, the spatial position (that is, the location) of a microscopic entity does not exist before the observational process, but is created by the very process of observation.

But that's not all. To say it all, the spatial position of a microscopic entity does not even exist after the observational process. Indeed, it is a property of an *ephemeral* nature [10].

At this point, perhaps you will ask: How can I understand the ephemerality and the incompatibility of quantum properties? Can I find deep analogies that can help me to better understand? Absolutely yes, and for this it is sufficient to love Italian spaghetti!

8. THE STRANGE PHYSICS OF SPAGHETTI

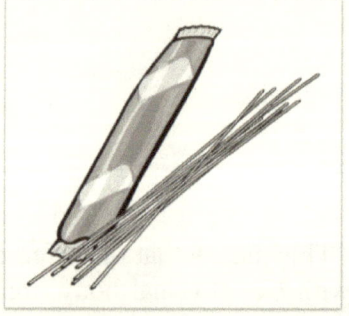

With the final sentence of the previous section I wanted to say exactly what I said: that some of the quantum mysteries that you have explored so far can be clarified by studying the strange physics of spaghetti, and more precisely of *raw spaghetti*!

For this, you will have to deal with the so-called (so to speak) *left-handedness* of spaghetti. Let me explain what it is. So, the physical system (or physical entity) you want to study is an uncooked spaghetti (preferably of a good brand). The measuring instrument you are going to use, in order to perform your observational experiments, that is, your measurements, is formed by your two hands.

The property that you want to observe, as I said, is the left-handedness. I know, almost surely you have never heard of the left-handedness of a spaghetti, but I will now explain what it is, by telling you how it is measured/observed.

You first have to grab the spaghetti with your two hands, as shown in Figure 27.

Figure 27. *The procedure for observing the left-handedness of a spaghetti requires to initially grab the spaghetti with the two hands.*

Then, you have to bend the spaghetti until it breaks; if the longest fragment remains in your left hand, then the left-handedness property is confirmed; otherwise, it is not confirmed. And if the spaghetti is already broken, you simply have to execute the test using the longest fragment.

In Figure 28, you can see a possible result of such process.

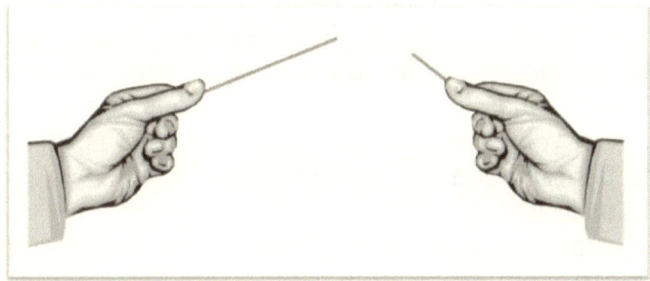

Figure 28. *The test has confirmed the left-handedness property of the spaghetti. In other words, the observation was successful.*

As you can observe, the test was successful, thus the left-handedness of the spaghetti was confirmed. In Figure 29, you can see another possible result of the left-handedness observational process.

This time the test was not successful, and the left-handedness of the spaghetti was not confirmed. Instead, it is the inverse property of left-handedness, which is the property of *right-handedness*, which was confirmed.

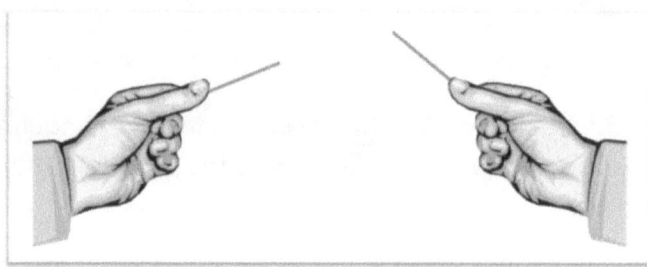

Figure 29. *The test has not confirmed the left-handedness property of the spaghetti. In other words, the observation was not successful.*

Fine, but let me now consider another property of the spaghetti, which I will simply call the *solidity*.

So, the physical entity to be measured is once more an uncooked spaghetti. The measuring instrument is this time only one of your hands, combined with the floor of your kitchen. Again, to let you know what the solidity property is, all I have to do is to describe you the experimental observational protocol, which is as follows.

You first have to hold the spaghetti in your hand, as indicated in Figure 30.

Then, you have to let it fall from your hand to the floor, from a height of about one meter; if it doesn't break, as in Figure 31, the solidity property is confirmed; if it breaks, as in Figure 32, the solidity property is invalidated, whereas it is the inverse property of solidity, *fragility*, which is confirmed.

And if the spaghetti is already broken, you simply do the experiment using the longest fragment.

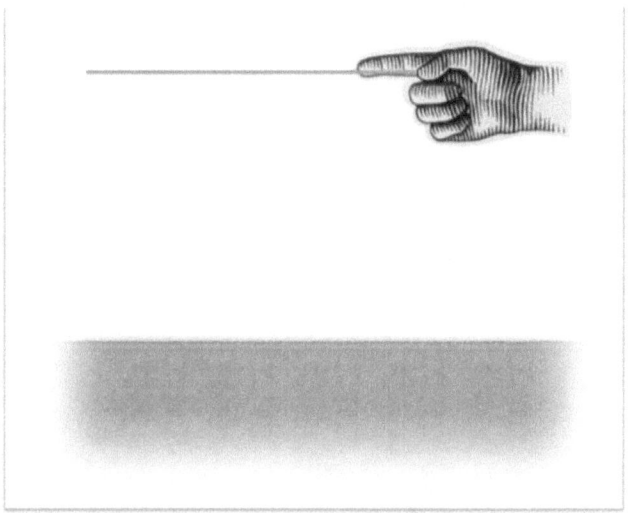

Figure 30 The procedure for observing the solidity of a spaghetti requires to initially hold the spaghetti in one of your hands.

Figure 31. The successful outcome of the solidity test.

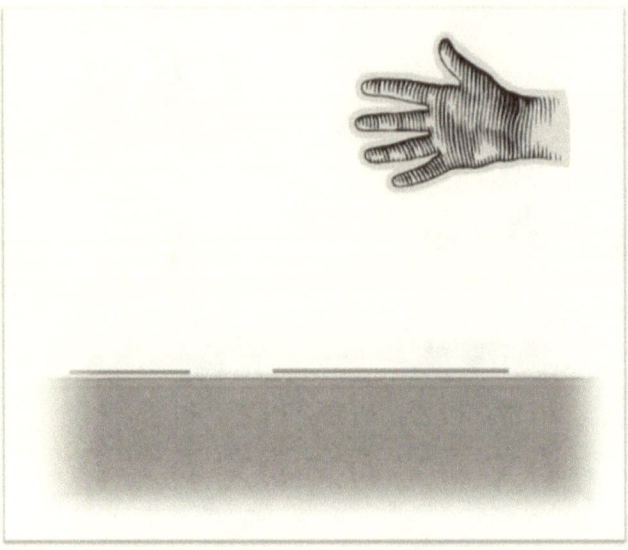

Figure 32. *The unsuccessful outcome of the solidity test, confirming the inverse property of fragility.*

Very well, if you have thought carefully, you may have understood that:

Left-handedness and solidity are ephemeral properties!

Indeed, consider the case of the left-handedness. Suppose that you have just made the observation and that the test was successful. At that precise moment, when you are still holding the two fragments of spaghetti in your hands, you can certainly say that the spaghetti actually possesses the left-handedness property. But as soon as you let go of the two fragments, that same left-handedness goes back to be a property which is only *potential*.

In fact, once the *relation* between the fragments of the spaghetti and the two hands of the experimenter is lost, it is no longer possible to affirm that the spaghetti is left-handed [13].

This because to observe again the left-handedness, you have to repeat the test, using the longer fragment, but nothing a priori guarantees you that its outcome will be again successful.

I hope you will appreciate the fundamental difference between the observation of a property such as the burnability of a wooden cube, and the left-handedness of a spaghetti. For the burnability, you were perfectly able to predict the outcome of the test, without any need to perform it. On the other hand, for left-handedness you are no longer in such situation.

Of course, you could argue that to make a reliable prediction you need to have all the necessary information, and for this carefully study in advance, maybe under a microscope (see Figure 33), all the characteristics of the spaghetti in question, perhaps also asking precise information from the manufacturer about the manufacturing method.

Figure 33. *Studying spaghetti under a microscope does not make it easier to predict the outcome of the left-handedness (or solidity) test.*

But will this really help you to predict in advance the outcome of the left-handedness (or solidity) test? If you attentively consider the way in which the left-handedness property (or the inverse property of right-handedness) is tested, you will easily convince yourself that:

Even with a complete knowledge of the spaghetti, up to the level of its molecular structure, you will never be able to predict with certainty the outcome of the observational test.

This is not because you would lack some essential information about the spaghetti as such (that is, about its actual *state*), but because you remain totally clueless about how exactly the process of the observation of the property will took place.

The outcome of the test will in fact depend on a number of variables that remain *hidden* to us, in the sense that are totally *outside of our control*, such as the subtle vibrations of your hands while you act on the spaghetti, its specific orientation, the variable pressure exerted by your fingers, the rapidity with which you bend it in order to break it, and so forth.

It is the combination of all these *hidden variables*, and those associated with the state of the spaghetti, which will determine, in an extremely complex way, in which point(s) the spaghetti is going to break, thus producing the final outcome of the test.

In other words, despite possibly having a complete knowledge of the state of the spaghetti, you will have no way to predict the effects of the innumerable fluctuations in the interaction between the spaghetti and the instrument of observation, made by your two hands; fluctuations that ultimately will determine the exact breaking point(s) of the spaghetti, and therefore either its left-handedness or right-handedness.

The logical conclusion of all this is that the property of the spaghetti of being left-handed or right- handed cannot be predicted in advance with certainty, also when you possess a complete knowledge of the state of the spaghetti.

And according to the EPR-PA criterion of reality, this means that these properties cannot be considered to be genuine elements of reality, i.e., they do not exist. Or, rather, their existence is only *potential*, in the sense that although they do not exist (they are not actual) in a given moment, they may nevertheless exist (be actualized) at a later instant. This is exactly what can happen during an observational process:

The left-handedness (or solidity) property is not <u>discovered</u> during its observational test, but possibly literally <u>created</u> by it. Before the test, it was not existing, but by means of the test it can be brought into existence, although only in an ephemeral way.

To sum up:
Left-handedness and solidity are both ephemeral properties, in

the sense that they are potential properties that are possibly created/actualized during an observation, by the observation itself, in a way that cannot be predicted in advance.

Also, they cease to be actual at the precise moment when the specific relation between the measuring instrument and the physical entity is severed.

Left-handedness and solidity, however, are also properties that are mutually incompatible. In fact, the observation of the left-handedness considerably increases the probability that a subsequent test of the solidity property will give a successful outcome, as is clear from the fact that the shortest the spaghetti the less easily it will break, when falling to the floor.

This means that, in general, performing first the test of left-handedness, then afterwards the test of solidity, for example on a large number of different spaghettis, the statistics of outcomes you obtain will differ significantly from that obtained by first performing the solidity test and then the left-handedness one.

This paradigmatic example of the spaghetti [10] reveals in an incontrovertible way that:

Incompatibility and ephemerality are independent notions.

Indeed, as Aerts' piece of wood example shows [6], two properties can be incompatible and nevertheless stably exist, at the same time; but as the spaghetti example also shows, two properties can be incompatible and only ephemerally exist, when actualized by their experimental test.

This indicates that the ephemeral character of a property has more to do with the way the property itself is defined (i.e., the way it is tested, in a practical way) than the fact that it may or not entertain incompatibility relations with other properties.

The above remark is particularly relevant in view of the fact that in our reasoning to deduce the non-spatiality of microscopic entities, the HUP (the existence of a relation of incompatibility between position and velocity) was used as a main ingredient.

Therefore, one could be tempted to conclude that it is the very existence of such an experimental incompatibility which is at the origin of the observed non-spatiality of the microscopic entities.

Considering the piece of wood example, we see however that incompatibility is not a sufficient condition for non-spatiality, and considering the spaghetti example, we also see that incompatibility is neither a necessary condition for it, as is clear from the fact that the ephemeral character of the left-handedness and solidity properties is built-in in the very definition of them, independently of the compatible or incompatible nature of their relation with other properties.

That being said, I hope I have not lost you in all these conceptual subtleties. What is really important to highlight here is that the left handedness and solidity of a spaghetti, like the position and velocity of a microscopic entity, are *non-ordinary* properties, in the sense that they are *non-classical* properties, non-spatiality being just an aspect of such non-classicality.

But perhaps you are now wondering: what exactly are classical properties? Well, simply, classical properties are properties obeying the so-called *classical prejudice* [9], stating that:

Classical prejudice: the outcome of an observational test is always a priori certain (predetermined), i.e., always predictable in advance, at least in principle.

But the classical prejudice has a very limited validity, being based on the wrong assumption that the interaction between the instrument of observation and the observed system/entity always takes place in a predeterminable way.

The observational tests of burnability and floatability of the wooden cube are certainly in accordance with the classical prejudice. But the observational tests of left-handedness and solidity of the raw spaghetti certainly invalidate the classical prejudice, in the same way it is invalidated by the observation of the position and velocity of a microscopic entity.

Very well. Let me now summarize the most important points of our investigation:

- We have seen that HUP expresses the experimental incompatibility of certain properties associated with microscopic entities, like position and velocity.
- We have also highlighted that experimental incompatibility is a widespread phenomenon, which manifests also with macroscopic bodies, and not only with microscopic ones.

- Moreover, and contrary to what one might initially believe, we have shown that it is perfectly possible to jointly test incompatible properties, by means of the so-called product tests.
- We have then highlighted the content of the EPR reality criterion, and its more complete EPR-PA version by Constantin Piron and Diederik Aerts, affirming that existence and predictability are intimately related notions (in the sense that, in ultimate analysis, a property is a state of prediction).
- Next, using in combination the HUP and the EPR-PA criterion, we have deduced the non- spatiality of microscopic entities, showing that to have a position is an ephemeral property of a microscopic entity, not stably possessed by it.
- We have then seen that ephemerality can also manifest in macroscopic bodies, and that incompatibility and ephemerality are independent notions.
- Finally, we realized that wooden cubes and uncooked spaghetti can be of great help in under- standing (and in part demystifying) some of the mysteries of quantum physics.

Of course, much more should be said to elucidate all these conceptually profound and subtle aspects of our physical theories. In particular, much should be added concerning the puzzling non-spatiality of quantum entities, often indicated by physicists with the less appropriate term of *non-locality* (as the latter implicitly suggests that the entity would still remain stably present within our spatial theater, although in a sort of spatially widespread condition).

9. THE FRIENDSHIP SPACE

I would like to conclude this presentation with a metaphor proposed in 1990 by *Diederik Aerts* [14], as an attempt to invent a world of entities where their spatial condition emerges from a different underlying reality, whose spatiality is of a different kind. This world of entities is, as I'm going to explain, a world living within a space of friendship.

More precisely, the entities considered are we human beings in a distant future, and the world of entities is our terrestrial human society. The interaction taken into account is the one of *friendship*, and the hypothesis is that the human society was able to survive by managing to totally eliminate enmity (negative friendship) and by making friendship always something *reciprocal*.

Let me explain what this means in more precise terms. If you denote by $ad(X, Y)$ the function that determines the affective distance that person X feels for person Y, and if $ad(Y, X)$ denotes the affective distance that person Y feels for person X, then reciprocity simply means that these two distances are identical or, to say it in more technical terms, that the "ad" function is *symmetrical*:

$$ad(X, Y) = ad(Y, X)$$

The *absence of enmity*, on the other hand, means that the function ad is always positive, as it should be the case for a distance worth of the name:

$$ad(X, Y) \geq 0$$

Let me consider now the ordinary physical space in which we humans live today. In this space there are people, for example, a boy, who I will call X, and a girl, who will call Y.

Between these two guys there is a *physical distance*, which I will denote "pd," which is the distance usually considered between the ordinary physical spatial objects, also a symmetrical function:

$$pd(X, Y) = pd(Y, X)$$

The boy X and the girl Y do not only exist in the ordinary physical space, but also in the space of friendship, and in this other space X and Y are not separated by a physical distance "pd," but by an affective distance "ad," which for example we can consider to be much smaller than the physical one, as is clear that the mutual friendship between two people does not depend on how distant they are in physical terms.

Consider now a third individual, who I will call Z, whose physical distance with Y is smaller than that between X and Y, and whose distance in the friendship space is larger than that between X and Y, for reasons that you can easily guess by looking at Figure 34.

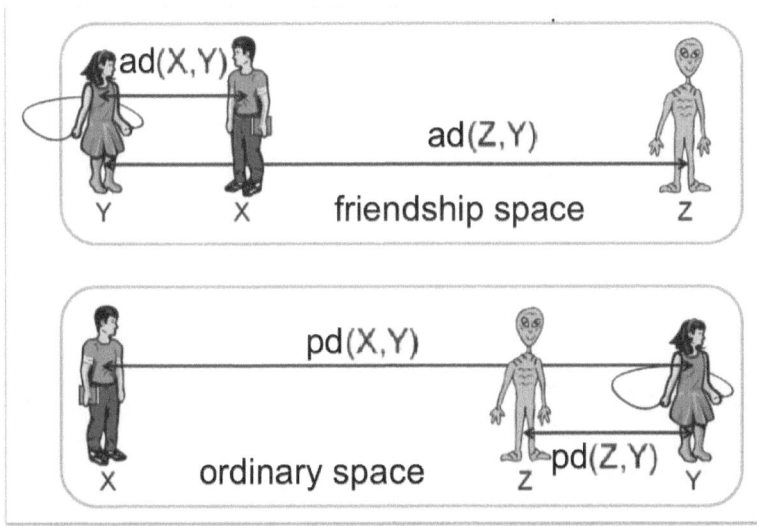

Figure 34. *Distances in the physical space and in the friendship space follow different logics: nearby objects in the physical space can be pretty much far away in the friendship space, and vice versa.*

Very well, up to here I have simply pointed out that the distances in the physical space and in the friendship space are not necessarily in correspondence with each other. Consider now the fact that in our human society, as time passes by, different subgroups of people will *emerge*, bound by specific affinities, which for simplicity I will consider to only be *affinities of an affective kind*. This emergence is responsible for a *structuring of the friendship interaction.*

Try to observe this phenomenon from the double perspective of the physical space and friendship space. In Figure 35, different persons (including X, Y and Z) are represented (for simplicity as simple dots). As you can see, although the different individuals are rather scattered in the ordinary physical space, they present themselves in a much more organized way in the friendship space.

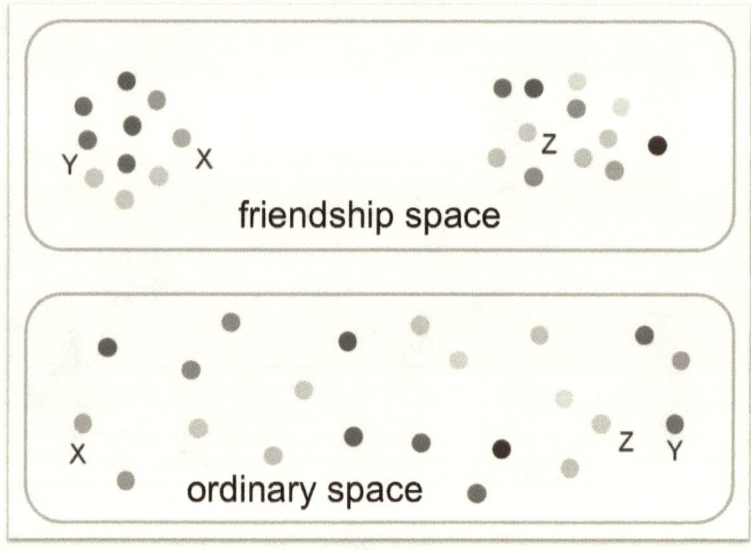

Figure 35. *The same individuals are represented in the friendship space (above) and in the ordinary physical space (below).*

More precisely, as made explicit in Figure 36, people end up organizing themselves into macrostructures. For simplicity, in Figure 36 I have only evidenced two of them, denoting them A and B, which are located at a certain (affective) distance $ad(A, B)$ from each other.

So, in the "old" physical space of the surface of planet earth, people live pretty much mixed together, but in the "new" structured space of friendship they are organized within specific *macrostructures*. To fix ideas, we can think of families, associations, interest groups, sects, etc. In other words, as time goes on, the friendship space will, little by little, becomes a perfectly structured *macrofriendship space*.

Consider now an additional individual K, and suppose that this individual, although at a given moment s/he is present in the ordinary physical space, s/he has not yet established a specific relationship with one of the macrostructures present in the macrofriendship space (see Figure 36).

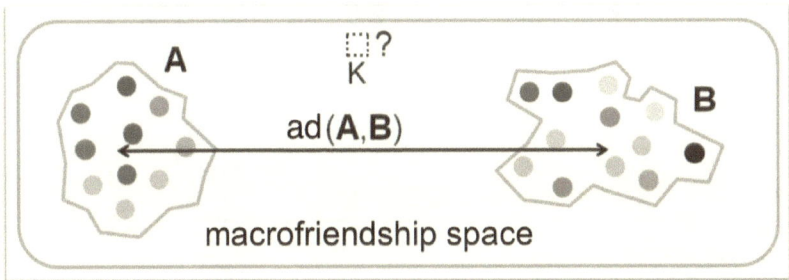

Figure 36. *Two macrostructures, A and B, in the space of macrofriendship, separated by a distance ad(A,B). The individual entity K, not belonging to A or B, is in a state of superposition with respect to these macrostructures, hence does not belong to the macrofriendship space.*

Suppose that after a very long time, humans of the future have totally forgotten about their original *Euclidean space*, associated to the surface of planet earth, as well as about the first version of their friendship space, when it was not yet structured into well-defined (affective) macro structures.

Then, in the fully structured space of macrofriendship, a single affectively isolated individual cannot have any localization, that is, a specific position, as having a well-defined localization in the macrofriendship space means to belong to a specific structure of affinity, in the present case either A or B.

The individual K, from the viewpoint of the macrofriendship space, is therefore a typical *non-spatial* entity, not present in actual terms in that space (see Figure 36).

Suppose however that as a result of a (non-spatial) interaction with the existing macrostructures, it comes the time when, for reasons that we do not need to specify here, K decides (or is forced) to choose to belong to either A or B.

Before that this happens, we can say that K is in a quantum-like *superposition state*, with respect to these two possibilities, and that at the exact moment s/he chooses to which macrostructure s/he belongs, K suddenly acquires (collapses to) a specific location in the macro-friendship space, becoming for example an integral part of the macrostructure B (see Figure 37).

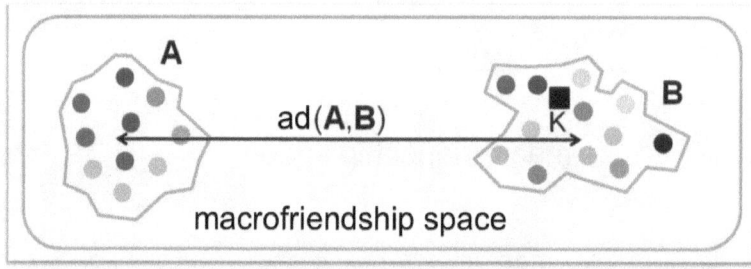

Figure 37. *The individual K, by choosing which affective macrostructure to belong to, passes from a non-spatial superposition state to a "collapsed" localized spatial state, relative to the space of macrofriendship.*

It is interesting to note that this process, during which a specific localization for the elementary entity K is suddenly created, in the macrofriendship space, through its interaction with the macrostructures A and B, is a typical *creation process*, which is reminiscent of the creation of a position for an electron, when it interacts with measuring instruments, which are precisely structures of a macroscopic kind, formed by a huge number of elementary microscopic entities, organized together.

For sure, the above is just a metaphor [14], though a very profound and enlightening one, which is certainly good to let it settle in your mind and on which it can be advantageous to further meditate. In fact, I believe the time has come to conclude this already quite long exposition.

10. READING SUGGESTIONS

If you have been interested in what you have read, and would like to deepen your reflection, here are some further reading suggestions.

Let me start with some texts that are readable also by those who are not experts in quantum physics. I will also then indicate a couple of articles for readers with some technical knowledge, wanting to dig into the formalism behind some of my explanations.

The fact that quantum (or quantum-like) measurements as also processes of *creation*, and not only as processes of *discovery*, is sometimes referred to as an *observer effect*, where the term generally refers to the possibility that an observation may affect the properties of what is observed. Examples and illustrations of such observer (creation) effect can be found in [15, 16, 13, 17].

The left-handedness (or solidity) property of spaghetti subtends a possible interpretation of quantum measurements called the *hidden-measurement interpretation* (HMI), which recently gave rise to a promising completion of the quantum formalism, known as the *extended Bloch representation* (EBR) of quantum mechanics.

A highly accessible introduction to the HMI and the EBR can be found in [18, 19, 20]; see also the video [21], where some nice computer animations of the unfolding of quantum measurements with two, three and four possible outcomes can be found.

For those readers who have fully mastered the quantum formalism, a more technical reading about the HMI and the EBR, containing all the mathematical details, can be found in [22, 23].

Finally, regarding the possibility of considering the micro-entities as non-spatial entities, it is important to say that there are many different ways to reach such conclusion.

In addition to my reasoning using the equations of motion, one can for instance analyze the remarkable experiments conducted in *neutron interferometry*, for instance those observing the 4π-periodicity of a neutron's spinor wave function, which one can be made to interfere with itself [24, 25, 12, 26].

And speaking of *spin*, it is possible to show that *spin eigenstates* cannot in general be associated with directions in the Euclidean space, but only with generalized directions in the *Blochean space* [27].

Non-spatiality can also be deduced by considering the *permanence time* of micro-entities in certain regions of space, which again can be shown to be incompatible with the very notion of a spatial trajectory [28].

To conclude, let me also point out a fascinating interpretation of quantum mechanics that was introduced by Diederik Aerts some years ago, and is currently under development, known as the *conceptuality interpretation*.

According to it, quantum entities would be non-spatial simply because they would be *conceptual* (*abstract*) *entities*, interacting among them and with the measuring apparatuses in ways that are analogous to how human concepts combine with each other in our linguistic constructions and interact with human minds. This not because human concepts and the microscopic physical entities would be the same kind of entities, but because they would share the same *conceptual nature*, similarly to how *sound waves* and *electromagnetic waves*, although very different entities, can nevertheless share the same *undulatory nature*.

A good place to start, to learn more about this truly fascinating interpretation, is the recent review article [29], and of course the references cited therein. A more concise video version of the article is also available on YouTube [30], with the presentation I gave at the Symposium "Worlds of Entanglement," organized by the Centre Leo Apostel for Interdisciplinary Studies and which took place at the Free University of Brussels (VUB), on September 29-30, 2017.

REFERENCES

[1] Sassoli de Bianchi, M. (2012, Apr. 5). "Principio di Heisenberg e Non-spazialità (Non-località) Quantistica" [Video file]. Available at: *https://youtu.be/nN3BWe4LanQ*.

[2] Sassoli de Bianchi, M. (2012, Aug. 27). "Heisenberg's

Uncertainty Principle and Quantum Non-Spatiality (Non-locality)" [Video file]. Available at: *https://youtu.be/9C3vtVADL1o*.

[3] Heisenberg, W. (1927). "Über den anschaulichen Inhalt der quantentheoretischen Kinematik und Mechanik," Zeitschrift fr Physik 43, 172–198.

[4] Compton, A. H. (1923). "A Quantum Theory of the Scattering of X-Rays by Light Elements," Physical Review 21, 483–502.

[5] Bohr, N. (1928). "The Quantum Postulate and the Recent Development of Atomic Theory," Nature 121, 580–590.

[6] Aerts, D. (1982). "Description of many physical entities without the paradoxes encountered in quantum mechanics," Foundations of Physics 12, 1131–1170.

[7] Einstein, A., Podolsky, B. & Rosen, N. (1935). "Can Quantum-Mechanical Description of Physical Reality Be Considered Complete?" Phys. Rev. 47, 777–780.

[8] Piron, C. (1976). Foundations of Quantum Physics, W. A. Benjamin Inc., Massachusetts.

[9] Piron, C. (1990). *Mécanique quantique. Bases et applications*, Presses polytechniques et universitaires romandes, Lausanne (Second corrected edition 1998).

[10] Sassoli de Bianchi, M. (2011). "Ephemeral properties and the illusion of microscopic particles," Found. of Sci, 16, 393–409.

[11] Laplace, P. S. (1825). *Essai philosophique sur les probabilités*; see for instance the English translation from the original French fifth ed. by Andrew I. Dale, in: *Sources in the History of Mathematics and Physical Sciences 13*, Springer-Verlag (New York, 1995).

[12] Aerts, D. (1999). "The stuff the world is made of: Physics and reality," in: D. Aerts, J. Broekaert and E. Mathijs (Eds.), Einstein Meets Magritte: An Interdisciplinary Reflection, 129–183. Dordrecht: Springer Netherlands.

[13] Sassoli de Bianchi, M. (2015). "God may not play dice, but human observers surely do," Found. Sci. 20, 77–105.

[14] Aerts, D. (1990). "An attempt to imagine parts of the reality of the micro-world," in: J. Mizerski et al., (Eds.), Problems in Quantum Physics II; Gdansk '89, pp. 3–25. World Scientific

Publishing Company, Singapore.

[15] Sassoli de Bianchi, M. (2013). "The observer effect," Foundations of Science 18, 213–243.

[16] Sassoli de Bianchi, M. (2013). Observer Effect. The Quantum Mystery Demystified, Adea edizioni, Sesto ed Uniti. (See also this volume).

[17] Sassoli de Bianchi, M. (2018). "Observer effect," The SAGE Encyclopedia of Educational Research, Measurement, and Evaluation (pp. 1172–1174). Edited by: Bruce B. Frey, SAGE Publications, Inc.

[18] Aerts, D. and Sassoli de Bianchi, M. (2015). "Many-Measurements or Many-Worlds? A Dialogue," Foundations of Science 20, 399–427.

[19] Aerts, D. and Sassoli de Bianchi, M. (2017). Universal Measurements. Singapore: World Scientific.

[20] Sassoli de Bianchi, M. (2018). "Using abstract elastic membranes to learn about quantum measurements," Foundations of Science. *https://doi.org/10.1007/s10699-018-9558-y*.

[21] Sassoli de Bianchi, M. and Sassoli de Bianchi, L. (2015, Feb. 11). "Solving the measurement problem" [Video file]. Available at: *https://youtu.be/Tk4MsAfC8vE*.

[22] Aerts, D. and Sassoli de Bianchi, M. (2014). The Extended Bloch Representation of Quantum Mechanics and the Hidden-Measurement Solution to the Measurement Problem. Annals of Physics (N. Y.) 351, 975–102. (See also the Erratum: Ann. Phys. 366, 197–198).

[23] Aerts, D. and Sassoli de Bianchi, M. (2016). The Extended Bloch Representation of quantum mechanics. Explaining superposition, interference and entanglement. J. Math. Phys. 57, 122110.

[24] Rauch, H., Zeilinger, A., Badurek, G., Wilfing, A., Bauspiess, W. & Bonse, U. (1975). "Verification of coherent spinor rotation of fermions," Phys. Lett. 54A, 425–427.

[25] Werner, S. A., Colella, R., Overhauser, A. W. & Eagen, C. F. (1975). "Observation of the Phase Shift of a Neutron Due to Precession in a Magnetic Field," Rev. Lett. 35, 1053.

[26] Sassoli de Bianchi, M. (2017). "Theoretical and conceptual analysis of the celebrated 4π-symmetry neutron interferometry experiments," Foundations of Science 22, 627–653.

[27] Aerts, D. & Sassoli de Bianchi, M. (2017c). "Do spins have directions?" Soft Computing 21, 1483–1504.

[28] Sassoli de Bianchi, M. (2012). "From permanence to total availability: a quantum conceptual upgrade," Foundations of Science 17, 223–244.

[29] Aerts, D., Sassoli de Bianchi, M., Sozzo, S. and Veloz, T. (2018). "On the Conceptuality interpretation of Quantum and Relativity Theories," Foundations of Science. *https://doi.org/10.1007/s10699-018-9557-z.*

[30] Sassoli de Bianchi, M. and Sassoli de Bianchi, L. (2017, Nov 2). "On Diederik Aerts Conceptuality Interpretation of Quantum and Relativity Theories" [Video file]. Available at: *https://youtu. be/-SteQN1A33M.*

autoricerca.com

ON THE QUANTUM "SELF-TELEPORTATION" PROBABILITY OF A HUMAN BODY

Massimiliano Sassoli de Bianchi

ABSTRACT. The probability of quantum relocation of a human body, at a given distance, is estimated using two different methods, giving comparable results. Not only the obtained values for the probabilities are inconceivably small, but assumptions of a sci-fi nature are also necessary to ensure that they are not identical to zero. The notions of 'non-spatiality' and 'superselection rule' are also briefly discussed.

autoricerca.com

1. INTRODUCTION

Recently a science fiction writer asked me the following question:[1]

What is the probability for an individual to suddenly vanish from one place and, one second after, reappear in another predetermined place, tens of kilometers away, according to the laws of quantum physics?

He also told me that a famous physicist (he had forgotten the name) used to pose such question to his students, so that it necessarily had to be a simple textbook problem. His interest in this question was that the protagonist of his story had to take advantage of this probability, no matter how infinitesimal (he was equipped with a futurist amplifier of probabilities), to "transfer" all of a sudden his body to a considerable spatial distance.

Inspired by his curious quiz, which only apparently is a textbook one, I will try in this article to offer a few elements of clarification about some important concepts of quantum physics, in particular the concept of non-spatiality, which I will illustrate by means of a simple metaphor. I will also provide two different estimates of the teleportation probability in question, on the basis of a number of simplifying assumptions, some of which will necessarily be of a sci-fi nature. Despite these assumptions, the values I will obtain are so small that they are almost impossible to conceive.

Not to create any misunderstanding, let me assert very clearly, from the beginning, that the probability that in normal conditions an individual would disappear from one place and be teleported to another place is, according to the today known laws of quantum physics (and the author's personal understanding of them), exactly equal to zero! In fact, as I will explain, and until evidence to the contrary, ordinary

---

[1] The author in question is *Marco Giacomantonio*, who I thank for the stimulating question.

macroscopic bodies (such as our human bodies), in standard environmental conditions (for instance of temperature and pressure), do not obey the quantum laws. But before proceeding in my discussion, I have to face a little problem of terminology.

The term "teleportation" is used in quantum physics to denote a very specific class of phenomena that have nothing to do with the nature of the question addressed to me by the science fiction writer [1]. These phenomena describe the possibility of carrying information from one place to another, in ways that allow the construction of an exact duplicate of a given physical entity. This construction can be obtained only on the condition that the system of origin (i.e., the system that is to be duplicated) is altered, if not destroyed, in the process, since a well-known theorem, called the no-cloning theorem, forbids to create a perfect clone of a quantum entity [2, 3].

Apart from this difficulty, the quantum teleportation, understood in the usual sense mentioned above, requires the preparation of special pairs of non-separated (entangled) systems that have to connect the two spatial regions between which the teleportation is to be produced, which for this reason is also called entanglement-assisted teleportation. In other words, this form of quantum teleportation requires the presence of technological apparatuses tailored to the specificities of the entities to be teleported, and the execution of a series of operations that will produce their destruction and reconstitution in the place of destination. This is not a spontaneous process, associated with probabilities, but a determinative process, which requires a specific technology to be implemented.

Let me add that in the quantum teleportation only the information about the entity is transported, so as to allow its reconstruction, while nothing material is actually moved (except the carriers of information along an ordinary communication channel); and even though experiments of entanglement-assisted teleportation have already been successfully carried out (the current record is a teleportation over a distance of 143 km, between the Canary Islands of La Palma and Tenerife [4]), these remain so far limited to individual microscopic entities and finite-dimensional physical observables, such as the polarization of a photon, or the spin of an electron.

Having said that, and in order to avoid misunderstandings, I will use in the following the term "quantum self- teleportation," or more simply self-teleportation, to designate a hypothetical process of spatial relocation of a physical entity, to distinguish it from the aforementioned quantum teleportation assisted by entanglement. As previously emphasized, self-teleportation does not seem to be possible for macroscopic bodies, in standard conditions, as they only obey the laws of classical physics. Therefore, some additional sci-fi-like assumptions will be needed to explore this possibility and provide an estimate of the probability of such event, for a macroscopic entity like the body of a human being.

2. NON-SPATIALITY

I will start by explaining a little better why a macroscopic body cannot behave like a microscopic entity. It is important to observe that macroscopic bodies, such as human bodies, or whatever ordinary objects, like rocks, grains of sand, etc., are *spatial entities*. This means that they evolve while remaining within the so-called 3-dimensional Euclidean space. To clarify what I mean by this, I will use a simple metaphor.

Imagine a swimmer in a pool. The pool's water corresponds to the 3-dimensional physical space, and the swimmer in it represents a macroscopic entity. If she wants to move from one point to another of the pool, that is, from one point in space to another point in space, she can only do so by swimming, and of course, due to the viscosity of water, the speed of her movement will be limited: she will not be able to exceed a determined maximum speed, which we can assume to be, say, of $2 \ m/s$. Thus, if we assume that the swimmer is located near the trampoline, and she wants to reach a point located at the center of the pool, say $10 \ m$ away, this will take her $5 \ s$, if she can travel at the maximum possible speed.

Imagine now a child on the trampoline. In this metaphor the

child is a microscopic entity, located outside of the pool's water, that is, outside the ordinary 3-dimensional physical space. Indeed, microscopic entities, when not organized into macroscopic aggregates, or when not interacting with macroscopic entities, are typically non-spatial entities [5–11], not belonging to the water of the pool. The child, as a non-spatial entity, "moves" through another "space," which in a sense is adherent to our physical space, and which in our metaphor is represented by the layer of air above the pool; and since the viscosity of air is lower than that of water, he will be able to do so with greater effective speed than the swimmer.

Suppose that the limit speed in the air, for the child, is of 10 m/s, and that he is actually running at that speed on the trampoline, while in the process of diving. He will then be able to pass from the region of the trampoline to the region of the center of the pool in about 1 s, which is something the swimmer is obviously unable to do. The interesting thing is that from the perspective of the swimmer, it is as if the diving child would appear out of nowhere in the middle of the pool, because he was not moving through the water, as the swimmer is forced to do, but through the air, which corresponds to a different layer of reality, of a non-ordinary kind, that we cannot directly perceive using our ordinary perceptual tools.

I hope it is clear to everyone that a swimmer who is immersed in the water of the pool will never be able to move from one point to another as a diver (who is outside of it) can do. Similarly, a macroscopic body (the swimmer in our metaphor), being forced to move while remaining in the 3-dimensional physical space, will never be able to mimic the behavior of a microscopic entity, which instead is almost always outside of it (unless of course we would find a way to bring it out from that water, which is the sci-fi hypothesis we will have later on to consider).

There are several ways to infer the mysterious non-spatiality of the microscopic entities. The simplest is to take seriously the uncertainty principle of Heisenberg. In fact, according to it, it is not possible to simultaneously determine both the position and momentum of a microscopic entity. Therefore, it is not possible to solve the equations of motion (which require as an input both quantities), and as a consequence it is not possible to determine the spatial trajectory of the entity in question. This impossibility does not arise from the fact that we would lack some crucial information about the state of the entity (as the *no-go theorems* about hidden-variable theories illustrate [12–17]), which if we would possess would allow us to determine its trajectory: it is an impossibility of a fundamental, irreducible nature, which forces us to acknowledge that such trajectory in space does not exist, and since it does not exists, we must also abandon the idea that a microscopic entity would be always present in the 3-dimensional space.

In the words of the previous metaphor, a microscopic entity is essentially a diver, not a swimmer, and if you look for a diver you will find him almost always on the trampoline, or in the air, and not in the water. On the other hand, the human body, which is a macroscopic entity, is a genuine spatial entity, that is, a swimmer, not a diver, who cannot disappear from space as if by magic, only to reappear in another region of the same; certainly not in normal conditions, and according to the known laws of physics.

Let me add a further terminological clarification. In the scientific literature the term non-spatiality is much less used than that of *non-locality*. However, both terms express the same idea. In fact, all that is stably present in our three-dimensional space is necessarily local, that is, locally present in it, in actual and not in potential terms (also an extended object, like a cloud, is a local object, as it possesses local actual properties). Therefore, what is not present in a local sense is in fact not present at all (which doesn't mean it doesn't exist), and consequently the concepts of non-locality and non-spatiality are intimately related.

Now, as I tried to illustrate with the metaphor of the pool, reality is layered, and one of these layers is that in which the microscopic entities live: it is a layer that could be called *prespatial* (and which in a sense is also *pretemporal*). In adherence to this prespatial layer (represented in the metaphor as the layer of air above the pool), lies our ordinary spatial layer (the water in the pool), inside which macroscopic entities usually evolve, like the objects of our daily lives, and our human bodies.

Of course, the pool metaphor is only to be understood as a very crude allegorical simplification. The non-spatial or prespatial layer is a non-ordinary reality whose dimensionality is much higher than the three dimensions of our ordinary space, or the four dimensions of spacetime, and in general it could even be considered to be infinite-dimensional (as infinite-dimensional is in general the Hilbert state-space of a quantum entity). This cannot be represented in the too simple pool metaphor, in which the dimensionality of the air region above the pool, and of the water region inside the pool, is the same. Also, the region of contact between the spatial and prespatial layers is much more articulate and intricate than what the metaphor suggests, and certainly the non-spatial (or pre-spatial) entities cannot be represented as simple corpuscular entities.

3. WAVE-PACKET SPREADING

The problem that we need first to consider is the evolution of the *probability of presence* (in space) of a microscopic entity, such as a single atom, when it evolves freely, i.e., when no external forces or other entities (microscopic or macroscopic) interact with it (apart the measuring system).

The term "probability of presence" should be understood in the sense of the probability with which the microscopic entity in question lends itself to the creation of a spatial localization, in a given region of space R, at a given time t, through its interaction with a measuring apparatus [6, 8].

In quantum theory, this probability is given by the squared modulus $|\psi_t(x)|^2$ of the wave function (or wave packet) ψ_t (describing the state of the entity in question, at time t), integrated over the spatial region R, that is:

$$\mathcal{P}_t(R) = \int_R dx \, |\psi_t(x)|^2$$

What I am now going to do is to estimate the width of such wave packet in the simplest case of a hydrogen atom, which is the first and simplest element of the famous Mendeleev's periodic table.

To determine the wave function of a hydrogen atom, it is useful to express the problem in the so-called variables of the *center of mass* and *relative movement*. In doing so, I will neglect for simplicity the description of the spins of the electron and proton. Without going into the details of this procedure, which can be found in any textbook of quantum mechanics, we can observe that due to this change of variables it is possible to transform the problem of two interacting bodies (electron + proton) into an effective, simpler problem, of two bodies that evolve independently of each other, and whose equations can therefore be solved separately.

The first body corresponds to the evolution of the center of mass of the system, and is equivalent to the evolution of a free entity (an entity evolving in the absence of any interaction) of total mass:

$$M = M_e + M_p$$

where M_e and M_p are the masses of the electron and proton, respectively. The second body corresponds instead to the evolution of an entity of (reduced) mass

$$\mu = \frac{M_e M_p}{M_e + M_p}$$

which moves in the presence of a Coulombian central force field.

The solutions of the Schrödinger equation associated with the first problem are the so-called plane waves, which cover a continuum of possible energies, from zero to infinity (we speak in this case of a continuous spectrum).

The solutions of the problem with the central force field are instead associated with discrete energy values, given by the well-known formula:

$$E_n = -\frac{E_I}{n^2}$$

where $n = 1, 2, \ldots$, and $E_I \approx 13.6\, eV \approx 22 \cdot 10^{-19}\, J$ is the *ionization energy of the hydrogen atom*. One speaks in this case of a discrete spectrum, corresponding to the known (emission and absorption) spectral lines, observed experimentally.

Now, as regards the possibility of acquiring different positions in space, what really matters is the movement of the center of mass of the hydrogen atom, which, as previously mentioned, evolves according to free evolution.

What we are interested in is to calculate the spatial spreading of the wave packet associated with the center of mass variable, since such spreading will provide us a good estimate of the probability of observing the hydrogen atom at a certain distance from the place where it was initially observed, say at time $t = 0\, s$.

The spatial spreading of the wave packet at time t can be estimated by calculating the so-called *standard deviation* ΔQ_t of the position observable Q (associated with the center of mass), which by definition is given by the square root:

$$\Delta Q_t = \sqrt{\langle Q^2 \rangle_t - \langle Q \rangle_t^2}$$

where "$\langle \ldots \rangle_t$" denotes the quantum average relative to the state of the center of mass entity. Using *Ehrenfest theorem*, with which one can calculate the average values of quantum observables, one can show (following a little long but not difficult calculation) that by judiciously choosing the origin of the time axis the spreading ΔQ_t of the center of mass wave packet, at time t, is given by:

$$\Delta Q_t = \sqrt{\frac{\Delta P_0^2}{M^2} \cdot t^2 + \Delta Q_0^2},$$

where ΔQ_0 is the spatial spreading at time $t = 0$, and ΔP_0 is the spreading with respect to the momentum observable P at time $t = 0$.

For the initial width ΔQ_0 of the packet, we can choose the typical value of the *Bohr radius* (which in the semi-classical model of the Danish physicist corresponds to the radius of the innermost electron), i.e., about $5.3 \cdot 10^{-11}\, m$ (0.53 *angstrom*).

For the value of ΔP_0 we can instead consider a dispersion which is compatible with the energy of the ground state of the hydrogen atom, i.e., such that:

$$\frac{\Delta P_0^2}{2\,M} \approx E_I$$

Considering that the total mass is: $M \approx M_p \approx 1.67\, kg$, we have $\Delta P_0 \approx 8.6 \cdot 10^{-23}\, J \cdot s/m$, which is compatible with *Heisenberg's principle*, as is clear that ($\hbar \approx 1.05 \cdot 10^{-34}$):

$$\Delta Q_0 \cdot \Delta P_0 \approx 45.6 \cdot 10^{-34}\, J \cdot s \approx 43 \cdot \hbar > \frac{\hbar}{2}$$

Inserting the above values into the previously obtained expression for ΔQ_t, and observing that the second term in the square root is negligible compared to the first, we thus obtain:

$$\Delta Q_t \approx t \cdot 5.1 \cdot 10^4\, m/s$$

That is:

$$t \approx 0.2 \cdot 10^{-4} \Delta Q_t\, s/m$$

This last expression tells us the time we roughly need to wait for the center of mass of the wave packet of the hydrogen atom to reach the spatial spreading ΔQ_t.

Let us consider some specific values. To obtain a spreading of 5 *km*, that is, of $5 \cdot 10^3\, m$, we have to wait about $10^{-1} s$, i.e., a tenth of a second. In 1 s, instead, the packet will have reached a width of about 50 *km*, while in 10 s its approximate width will be of 500 km, and so forth.

In other words, the *effective speed* with which the radial dimension of the center of mass wave packet grows, is approximately 50 km/s, that is 180,000 km/h, which is a speed of all respect, and corresponds, in our previous metaphor, to the maximum speed of the diver (from our ordinary spatial perspective this is however only a *potential* speed, and certainly not an actual speed, as it is not associated with a body moving through our ordinary space).

4. DISASSEMBLING THE BODY

Summarizing, for a hydrogen atom we have determined the approximate behavior of that part of the wave function which describes the potential spatial localization of its center of mass. In doing so, we have ignored for simplicity the relative motion between the proton of the nucleus and the orbital electron, as well as their spins.

More precisely, we have calculated how the width of the center of mass wave function varies over time in (configuration) space, due to the so-called phenomenon of the *spreading of the wave function*, which can be understood as being a consequence of Heisenberg's uncertainty principle.

What is important to understand is that the domain in which the wave function is sensibly different from zero corresponds to the spatial region within which the atom in question has a chance of being detected. So, if the hydrogen atom, at time $t = 0$, was localized in a sphere whose radius is approximately equal to the Bohr radius, i.e., $r_0 = 5.3 \cdot 10^{-11} \, m$ (that is, its probability of presence in that sphere, at time $t = 0 \, s$, is equal to 1), what we have determined is that after for example 1 s, that localization radius will have approximately grown to about: 50 $km = 5 \cdot 10^4 \, m$.

What we are interested in is to estimate the probability with which we can detect the atom not in any location of this macro-sphere of 50 km of radius, but in a predetermined sub-region of it. Indeed, if we later want to extrapolate our reasoning to an entire macroscopic structure, it is necessary that every atom forming the structure will re-locate in a very specific place in relation to all the other atoms of the structure, so as to reconstitute it in every detail. So, let us suppose that this sub-region corresponds to a micro-sphere whose radius is equal to the Bohr radius r_0.

To estimate the above probability, I will make an additional simplifying assumption. The wave function being not a constant function, the probability of presence will vary according to the location of the micro-sphere within the macro-sphere. However, since we are only interested in estimating a rough order of magnitude, we can assume that the wave function is a step-function, only taking two values: a constant non-zero value inside the macro-sphere, and a zero value outside of it.

With this simplification, we have everything we need to complete our estimation. For this, we have to remember that to calculate a probability of presence we have to integrate the squared module of the wave function over the region of interest.

Considering the step-function hypothesis, this means that the probability that we seek will be *proportional to the relative volume of the micro-sphere compared to the volume of the macro-sphere.*

More exactly, given that the volume of a sphere is proportional to its radius to the cube, we obtain for the probability p that the hydrogen atom in question will be detected, after 1 s, in a predetermined micro-sphere of radius r_0, within the macro-sphere of radius $r = 50\ km$, the following order of magnitude:

$$p_H \approx \frac{r_0^3}{r^3} \approx \left(\frac{5.3 \cdot 10^{-11} m}{5 \cdot 10^4\ m} \right)^3 \approx 10^{-45}$$

This is undoubtedly a very small number, with 45 zeros after the decimal point! And of course, we can easily do the same calculation for larger macro-spheres, i.e., waiting more time than just a second. For example, if we wait 10 s, the radius of the macro-sphere will increase by a further factor of 10 (from 50 km to 500 km), and consequently the estimated value of the probability p will decrease from 10^{-45} to 10^{-48}, i.e., by a factor of a thousand, and so on.

Now that we have obtained an estimate of the *probability of quantum self-teleportation* of a hydrogen atom from an initial micro-sphere to a given final micro-sphere, we must consider the case of an entire macroscopic body, like that of a human being of planet Earth, which we can assume to have a mass of 100 kg.

Here of course we have to face the already mentioned problem that a macroscopic body has the property of spatiality, and therefore cannot be conveniently described by a wave function (more will be said about this in the next section). But suppose that for some reasons, unknown to us, all the interatomic bonds suddenly cease to exist, so that in an instant all the atoms that form the human body in question become separate and independent from each other, bringing them back to the prespatial layer of our physical reality.

On the basis of this sci-fi hypothesis, each individual atom of the body structure can be conveniently described by the laws of quantum mechanics, and we can apply the previous calculation to each one of them. In fact, this is not really true, as is clear that all these atoms will constantly be bombarded by the countless entities present in the environment, in particular the thermal photons, so that we also need to assume that the sci-fi process of disassembly of the human body is able to induce a perfect isolation of the different atomic constituents from all the other entities (micro and macro) present in the environment.

To keep the discussion as simple as possible, we further assume that the body structure is constituted solely by hydrogen atoms, and since the mass of a hydrogen atom is about $1{,}67 \cdot 10^{-27} kg$, a human of 100 kg, if constituted only by hydrogen atoms, would contain approximately a number N of them given by:

$$N \approx \frac{100 \ kg}{1.67 \cdot 10^{-27} \ kg} \approx 4.2 \cdot 10^{28} \approx 10^{28}$$

Each of these atoms will have to individually re-locate in a specific region of space, to reconstitute the entire body structure, with no errors. Therefore, the (estimated) probability $\mathcal{P}_{body}$ of *self-teleportation of the overall body structure* will be given by the product of the self-teleportation probabilities $\mathcal{p}$ of every single atom contained in that structure.

If the body would be formed only by two atoms, that is, $N = 2$, the probability would be:

$$\mathcal{P}_{body} \approx \mathcal{p} \cdot \mathcal{p} = \mathcal{p}^2 \approx 10^{-45 \cdot 2} = 10^{-90}$$

With three atoms, the probability would become:

$$\mathcal{P}_{body} \approx \mathcal{p} \cdot \mathcal{p} \cdot \mathcal{p} = \mathcal{p}^3 \approx 10^{-45 \cdot 3} = 10^{-135}$$

Therefore, with $N = 10^{28}$ atoms, we obtain:

$$\mathcal{P}_{body} \approx \mathcal{p} \cdots \mathcal{p} = \mathcal{p}^N \approx 10^{-45 \cdot N} = 10^{-4.5 \cdot 10^{29}}$$

Let us reflect for a moment on the amazing infinitesimality of this number. To write it in decimal, non-scientific notation, we must use more than 10^{29} zeros, i.e., more than one hundred billion billion billion zeros!

If we write with a printer on paper ten zeros per second, to write the entire number will take us more than 10^{28} seconds, that is, more than 10^{21} years, which is about a hundred thousand billion times the assumed age of the known universe (according to today cosmological theories)!

In other terms, the value we have obtained for $\mathcal{P}_{body}$, although not strictly equal to zero, is nevertheless so small that we have no point of comparison to be able to understand it. Yet, we have probably overestimated it.

In fact, we have assumed that for a reason unknown to us

all the atoms of the human body will suddenly disassemble and become non-spatial entities, so allowing their individual wave packets to spread. But we have also neglected the problem of the relative motion between the different atomic constituents, equating the individual atoms to free elementary-like particles.

Furthermore, we have neglected the spin variables of the different atomic constituents. Also, we have assumed that the environment in which the different atomic components evolve, once disassembled, corresponds to an effective absolute vacuum, otherwise the associated wave-packets cannot be considered to evolve freely, and that each atom is able to evolve without interacting with all the others, before regaining a specific spatial location.

In addition to that, we have hypothesized that when the various atoms reappear in the relative positions they occupied before being disassembled, the entire macroscopic structure will be able to reconstitute, without any particular inconvenience. Taking into account all these assumptions would of course further reduce the value of $\mathcal{P}_{body}$, by a factor which is very difficult, if not impossible, to evaluate.

But that's not all. There is another "sci-fi miracle," which is implicit in our reasoning, perhaps even more amazing than that of the disassembly of the initial structure (which is possible to relax; see the next section). This second miracle has to do with the different hydrogen atoms being simultaneously "drawn" back into space. Let me explain. An elementary entity, such as a proton, an electron, or an entire hydrogen atom, spends most of its time in a non-spatial (non-local) condition, unless it is incorporated into a macroscopic structure. Now, although there is no consensus on this among physicists, many agree that a microscopic entity is unable to acquire a precise spatial localization spontaneously, as this can only be done by interacting with a macroscopic material structure, like for instance that forming a measurement apparatus.

Experimental physicists are undoubtedly able to build

detection apparatuses allowing microscopic entities to temporarily acquire a spatial localization, in specific places, and even though these apparatuses could in principle localize in space a certain number of microscopic entities at a time, as far as I know a device which can localize an entire macroscopic structure doesn't exist, and perhaps is not even conceivable.

The possibility remains, of course, that the process of spatial localization could occur even in the absence of macroscopic structures playing the role of detection devices, as is suggested in some interpretations of quantum theory, like the so-called objective collapse theories [18], the transactional interpretation of quantum physics [19], and others, the discussion of which, however, would go beyond the scope of the present article.

Among the factors that we have not taken into account in the estimation of p_{body}, there are of course also those that could slightly increase the value of the probability. For example, we have implicitly assumed that all the atomic components have to re-localize at exactly the same instant. However, nothing prevents us from admitting a small time-delay in the localization process of the individual atoms, which, if sufficiently small, may not affect the correct re-assembly of the entire body macrostructure. But it is unlikely that considerations of this kind would be able to significantly change the infinitesimality of p_{body}.

5. COOLING DOWN THE BODY

At this point some readers may rightly object that we don't really needed our "disassembling sci-fi hypothesis," as what generally makes a macroscopic object like our human body behave classically, i.e., spatially and locally, is just the fact that it is immersed in a thermal environment, i.e., that it is constantly subjected to the random collisions of countless microscopic entities, in particular photons, and that the overall effect of these innumerable interactions is that of producing its continuous "collapse into space," which would be essentially the reason why it would behave differently from a "pure" quantum entity, like an electron.

To use once more our metaphor, this bombardment is what would force the body to remain inside the water of the pool, preventing its owner from becoming a diver.

So, one could object that, to allow the body to quit the "spatial pool," and temporarily become a non-spatial entity, it would be sufficient to shield it from the external thermal environment, so that there would be no need for having it first disassembled into smaller atomic fragments and then recombined, which is the operation that apparently produced the inconceivable infinitesimality of the self-teleportation probability, as each of the 10^{28} fragments needed to re-localize in a predetermined place, within a sphere of $50\ km$ of radius.

This is a pertinent objection that I'm now going to explore. This objection, by the way, could appear to be in contradiction with what I have just stated above, at the end of the last section, regarding the lack of an apparatus that could objectify (spatialize) a whole macroscopic structure. If our standard terrestrial environment is able to keep a macroscopic body into space, then wouldn't be that same environment the measuring apparatus that is able to achieve the required goal of producing the collapse – the objectification – of an entire macroscopic object? If this is true, then it would be sufficient to isolate an ordinary object to obtain its automatic de-localization (i.e., its de-spatialization).

However, we cannot expect this to work, as the object is also in contact with another environment: its own *internal* one. If the body is sufficiently large, as is certainly the case of a human body (but also of a speck of dust, and of much smaller entities), then the mutual interactions of its constituents can also have an influence in determining its overall classical (spatial) versus quantum (non-spatial) behavior.

The reason for this is easy to explain. In the case of the hydrogen atom, we were able to separate the wave function relative to the center of mass from that associated with the relative motion.

In this way, the center of mass was described by a free evolving wave function. With a macroscopic body, we may want to do the same, i.e., to separate the wave function describing the center of mass from the contribution coming from the different movements of all its constituents, relative to that center and to each other. Here we can consider the ensemble of these constituents as an entity playing the role of a measuring apparatus with respect to the "center of mass entity," so that the latter would be constantly subjected to a measurement process, thus producing its classical behavior.

Therefore, to describe the center of mass by means of a free evolving wave function, the evolution of the body's center of mass needs to decouple from that of its internal degrees of freedom, and this can reasonably be done only if the body is cooled down to extremely low temperatures. How low? Well, we can say, remaining here necessarily vague, low enough to avoid any exchange of energy between the center of mass degree of freedom and the degrees of freedom associated with the internal relative movements [20].

In the previous section we have assumed that by some sci-fi action the body was all of a sudden disassembled (and each constituent isolated from one another, and the environment). This was an assumption of simplicity, as in this way we were able to use the well-known factorization of the wave-function of a two-body system, which of course is much harder to obtain in general for a macroscopic body. We can however replace the "disassembling sci-fi hypothesis" with the requirement that not only the body in question will have to evolve in a perfect vacuum (no thermal bombardment), but also that it will be cooled down instantaneously to temperatures almost equal to the absolute zero.

In other terms, we now replace the "disassembling sci-fi hypothesis" by a "freezing sci-fi hypothesis." The advantage is that cooling down a body seems an operation less impossible to achieve than disassembling it into its atomic fragments, without destroying them, but also, and more importantly for our idealized discussion, the entire structure of the body will be preserved in this way, which hopefully will increase the value of the self-teleportation probability. Of course, also in this case we will have to assume that the body is additionally isolated from the environmental thermal bombardment.

So, let us assume, as we did before, that the mass of the body is 10^2 kg. By assumption, since the internal and external environments have now been made totally silent, and cannot anymore play the role of generalized detector instruments with respect to the wave function of the body's center of mass, we can consider that the latter is described by a free evolving wave packet. To further simplify the discussion, we assume that at time $t = 0$ s, the wave packet is approximately Gaussian (this is a reasonable assumption, considering that the probability density of any non-Gaussian wave packet becomes approximately Gaussian as it spreads [22]), which means that the inequality in Heisenberg's uncertainty principle is approximately an equality:

$$\Delta Q_0 \cdot \Delta P_0 \approx \frac{\hbar}{2}$$

As we did with the hydrogen atom, we take the standard deviation of the center of mass position observable ΔQ_0 to be equal to the Bohr radius. Therefore:

$$\Delta P_0 \approx \frac{\hbar}{2\Delta Q_0} \approx \frac{1.05 \cdot 10^{-34} \, J \cdot s}{2 \cdot 5.3 \cdot 10^{-11} \, m} \approx 10^{-24} \, kg \cdot m \cdot s^{-1}$$

Inserting this value into the previously obtained expression for ΔQ_t, we find that the spreading ΔQ_t of the center of mass wave packet at time t is given by:

$$\Delta Q_t = \sqrt{(10^{-52} \, m^2 \cdot s^{-2})t^2 + 3 \cdot 10^{-21} m^2}$$

Now, the original question of the sci-fi writer was to have the body disappearing from one place and reappearing tens of kilometers away. Considering, as we did before, a distance of

$50\ km$, we have to set $\Delta Q_t = 50\ km = 5 \cdot 10^4\ m$ in the above equation. If we do so and solve for t, we find the value $t \approx 5 \cdot 10^{30}\ s$. Considering that one year corresponds to $3.154 \cdot 10^7\ s$, we obtain that the center of mass wave packet of the macroscopic body will reach a width of $50\ km$ after approx. $1.6 \cdot 10^{23}$ years, which is approximately ten million billion times the assumed age of the known universe!

Here we see an important difference between the wave packet spreading of a hydrogen atom, who was extremely fast, and the wave packet spreading of a macroscopic body, which is inconceivably slow. But to answer the question of the sci-fi writer, we certainly cannot wait so long, because he explicitly asked the self-teleportation to happen in a matter of seconds. Also, even in case we would accept to wait for so long, we may have a problem with the "expiration date" of our universe! And anyhow, without having to freeze the body and the environment, using any classical means of transport through space (the "swimming modality") would be in this case much more effective to travel the distance of $50\ km$.

On the other hand, we can also say that, although the width of the wave packet almost doesn't increase as time passes by, even if we perfectly confine the position of center of mass in a given small region of space, at time $t = 0$ s, a fraction of a second after its wave function will have acquired an infinite tail. The value of this tail will be infinitesimally small, but nevertheless different from zero. So, let us estimate this value, at a distance of $50\ km$, after exactly one second of free evolution.

The body's center of mass wave packet can be written as the product of three identical Gaussian factors:

$$|\psi_t(x)|^2 = |\psi_t(x_1)|^2 |\psi_t(x_2)|^2 |\psi_t(x_3)|^2$$

where ($i = 1,2,3$):

$$|\psi_t(x_i)|^2 = \sqrt{\frac{2}{\pi a^2}} \ \frac{e^{-\frac{2a^2(x_i-\frac{\hbar k_i}{M}t)^2}{a^4+\frac{4\hbar^2 t^2}{M^2}}}}{\sqrt{1+\frac{4\hbar^2 t^2}{M^2 a^4}}}$$

To evaluate $|\psi_t(x)|^2$ at time $t = 1$ s, we can set $k_1 = k_2 = k_3 = 0$ (the body is at rest at time $t = 0\,s$), $x_2 = x_3 = 0\,m$, $x_1 = 5 \cdot 10^4\,m$, $M = 10^2\,kg$, and $a = 2r_0 \approx 10^{-10}\,m$. This implies that $|\psi_t(x_2)|^2 = |\psi_t(x_3)|^2 \approx 10^{10}$, and $|\psi_t(x_1)|^2 = |\psi_t(x_2)|^2 \approx 10^{10} \cdot 10^{-5 \cdot 10^{29}}$.

Multiplying the probability density $|\psi_t(x)|^2$ by the volume of the micro-sphere of Bohr radius in which we want the center of mass to relocate, at time $t = 1\,s$, we thus obtain that:

$$\mathscr{P}_{body} \approx 10^{-5 \cdot 10^{29}} \approx 10^{-2.2 \cdot 10^{29}}$$

Comparing this value with that obtained in the previous section, we observe that have obtained a self-teleportation probability of the same order of magnitude.

This means that, quite surprisingly, even if we avoid the sci-fi procedure of disassembling the macroscopic body, which as we have seen was responsible for the inconceivable infinitesimality of the obtained probability, and replace it with a procedure of total internal freezing, thus preserving the structural integrity of the body, a similar inconceivably infinitesimal self-teleportation probability is obtained, this time because of the extreme slowness of the spreading of the macroscopic wave function and the extreme infinitesimality of its long-distance tails.

6. SUPERSELECTION RULES

Considering my above pessimistic analysis, I'll leave it to the science fiction writer the task of finding a convincing sci-fi solution, not violating too many physical laws at the same time, allowing the hero of his story to teleport himself and accomplish his mission, whatever it is. As for me, let me offer a final thought.

According to quantum theory, and the phenomenon of the spreading of the wave packet, a hydrogen atom, if left to evolve freely, will quickly acquire a truly gigantic size, apparently in contradiction with what is usually observed. Also, when considering the spectrum of energies of a hydrogen atom, in addition to the discrete energy values, associated with the relative electron-proton movement, we also have to consider the continuous energy values associated with the translational degrees of freedom of the center of mass. The spectrum of the total energy of the atom is thus given by the sum of these two energy spectra. But the sum of a discrete spectrum and a continuous spectrum produces a continuous spectrum, apparently in contradiction with the spectral lines experimentally observed.

In short, without further precautions, the application of the Schrödinger equation to the problem of the hydrogen atom does not allow to obtain results in agreement with the experimental observation, that is, in agreement with the fact that atoms do not usually possess macroscopic sizes, nor spectra of a continuous nature. To solve this problem, one can make use of the notion of *superselection rules* [21].

Rules of this kind restrict the physically realizable states, and when associated with a given observable, they prevent considering states that would be a superposition of states associated with different values of this observable, as these superpositions would be in disagreement with the experimental data. In other terms, the existence of superselection rules indicates that the structure of the state space is not strictly Hilbertian (as linearity would not apply for all states), but more general.

An example of superselection rule is that associated with the observable determining whether the infamous Schrödinger's cat is alive or dead. If ψ_A is the wavefunction describing the alive cat, and ψ_D the wave function describing the dead cat, then, as far as we know, the wave function $\psi = \psi_A + \psi_D$ obtained by superposing these two wave functions does not describe a physically realizable state. This means that there is a superselection rule on the "life observable" of the cat, which forbids the superposition of wave functions characterized by different values of this observable.

In the case of the hydrogen atom, if we want to obtain values for its energy spectrum in agreement with the experimental data, it is necessary to consider the position and momentum of its center of mass as variables of a classical kind, associated with superselection rules, and same thing if we want to correctly describe its observed non- macroscopic size. Of course, the reasons for this inhibition of quantum superpositions and the associated classical behavior of certain observables can be multiple, and will generally depend on the specificities of the environment in which the entity in question is immersed. So, determining what are the classical observables and what the quantum ones, in a given context, is a problem not necessarily easy to solve, and there is no unique recipe for this: in some contexts, certain observables will behave classically, while in other contexts they will behave quantum mechanically, and still in others their behavior will be semiclassical, or semi-quantum, that is, in between these two regimes.

But then, if the center of mass of the hydrogen atom is the expression of a superselection rule, goodbye self- teleportation! On the other hand, if we consider it as quantum observable, goodbye agreement with many experimental data. But as I said, to determine the classical or quantum nature of an observable it is necessary to take into account the specificities of the experimental context. When an atom is incorporated into a macroscopic material structure, or undergoes continuous interactions with countless microscopic entities and force fields present in the environment, it usually undergoes a process of de-synchronization of its wave function, able to transform certain quantum observables into classical ones. That's why, in the beginning of this article, I have argued that, strictly speaking, quantum self- teleportation would be impossible.

More precisely, it is impossible ($\mathcal{P}_{body} = 0$) if we consider the standard environment in which we humans evolve, which makes our bodies, and the objects with which we interact in our everyday life, classical entities.

What I'm here suggesting is that the classical or quantum nature of a physical entity is not an intrinsic feature of the same, but a contextual one: in some contexts, certain entities will behave as quantum entities (when subjected to certain observational processes), and in other contexts they will behave, instead, classically. These considerations open to an important reflection, which can be summarized in the following question:

Is the physical reality fundamentally quantum?

The majority of physicists seem to believe so, that is, to think that quantum theory would be more fundamental than classical theory, and that a classical behavior would always emerge from a quantum substrate, when certain circumstances are met. However, a different view is also possible. For instance, one can consider that our physical reality is neither classical nor quantum, but genuinely hybrid, that is, a complex combination of these two aspects.

In other words, the physical entities forming our reality would generally be quantum-like, i.e., they would be entities potentially manifesting both aspects, the classical and the quantum aspects, depending on their state and the nature of the experimental questions we address to them. According to this view, supported by some very general (operational) approaches to the foundations of physical theories, especially that of the so-called Geneva-Brussels School of Quantum Mechanics (nowadays mainly active in Belgium, at the *Center Leo Apostel*, led by the Belgian physicist *Diederik Aerts*), the classical regime and the quantum regime would correspond to very specific limit cases of more general situations [5, 6, 9, 23].

More precisely, the classical regime would be associated with experimental situations where all fluctuations can be controlled, so that all observational processes are predictable in advance. On the other hand, the quantum regime would be associated with experimental contexts in which the fluctuations are maximal (and uniform), so producing a situation of maximum lack of knowledge. In between these two regimes, intermediate, hybrid regimes can also exist, neither purely classical nor strictly quantum, that physicists have just begun to investigate and that seem to provide a more general and realistic model for the description of the countless physical entities in interaction with their multiple environments.

REFERENCES

[1] C. H. Bennett et al., "Teleporting an Unknown Quantum State via Dual Classical and Einstein-Podolsky-Rosen Channels," Phys. Rev. Lett., 70, 1895–1899 (1993).

[2] W. Wootters and W. Zurek, "A Single Quantum Cannot be Cloned," Nature 299: 802–803 (1982).

[3] D. Dieks, "Communication by EPR devices," Physics Letters A 92(6): 271–272 (1982).

[4] X. Ma, T. Herbst, T. Scheidl, D. Wang, S. Kropatschek, W. Naylor, B. Wittmann, A. Mech, J. Kofler, E. Anisimova, V. Makarov, T. Jennewein, R. Ursin and A. Zeilinger, "Quantum teleportation over 143 kilometres using active feed-forward, Nature 489, 269–273 (2012).

[5] D. Aerts, "The entity and modern physics: the creation discovery view of reality," In: *Interpreting Bodies: Classical and Quantum Objects in Modern Physics*, edited by Elena Castellani (pp. 223–257). Princeton Unversity Press, Princeton (1998).

[6] D. Aerts, "The Stuff the World is Made of: Physics and Reality," In: *The White Book of 'Einstein Meets Magritte'*, edited by Diederik Aerts, Jan Broekaert and Ernest Mathijs (pp. 129–183). Kluwer Academic Publishers, Dordrecht (1999).

[7] M. Sassoli de Bianchi, "Ephemeral Properties and the Illusion of Microscopic Particles," Foundations of Science 16,

393–409 (2011).

[8] M. Sassoli de Bianchi, "From Permanence to Total Availability: A Quantum Conceptual Upgrade," Foundations of Science 17, 223–244 (2012).

[9] M. Sassoli de Bianchi, "The δ-Quantum Machine, the k-Model, and the Non-ordinary Spatiality of Quantum Entities," Foundations of Science 18, 11–41 (2013).

[10] D. Aerts and M. Sassoli de Bianchi, "The extended Bloch representation of quantum mechanics and the hidden-measurement solution to the measurement problem," Annals of Physics 351, 975–1025 (2014).

[11] M. Sassoli de Bianchi, "God May Not Play Dice, But Human Observers Surely Do," Foundations of Science 20, 77–105 (2015).

[12] J. Von Neumann, "Grundlehren," Math. Wiss. XXXVIII (1932).

[13] J. S. Bell, "On the Problem of Hidden Variables in Quantum Mechanics," Rev. Mod. Phys. 38, 447–452 (1966).

[14] A. M. Gleason, "Measures on the closed subspaces of a Hilbert space," J. Math. Mech. 6, 885–893 (1957).

[15] J. M. Jauch and C. Piron, "Can hidden variables be excluded in quantum mechanics?," Helv. Phys. Acta 36, 827–837 (1963).

[16] S. Kochen and E. P. Specker, "The problem of hidden variables in quantum mechanics," J. Math. Mech. 17, 59–87 (1967).

[17] S. P. Gudder, "On Hidden-Variable Theories," J. Math. Phys 11, 431–436 (1970).

[18] G. C. Ghirardi, "Collapse Theories," The Stanford Encyclopedia of Philosophy (Winter 2011 Edition), Edward N. Zalta (ed.).
http://plato.stanford.edu/archives/win2011/entries/qm-collapse/.

[19] R. E. Kastner, *The Transactional Interpretation of Quantum Mechanics,* Cambridge University Press, Cambridge (2013).

[20] C. P. Sun, X. F. Liu, D. L. Zhou and S. X. Yu, "Localization of a macroscopic object induced by the

factorization of internal adiabatic motion," Eur. Phys. J. D 17, 85–92 (2001).

[21] R. F. Streater and A. S. Wightman, PCT, spin and statistics, and all that, W. A. Benjamin, Inc., New York (1964).

[22] K. Mita, "Dispersion of non-Gaussian free particle wave packets," Am. J. Phys. 75, 950–952 (2007).

[23] D. Aerts, "Quantum Theory and Human Perception of the Macro-World," Front. Psychol. 5, article 554 (2014).

autoricerca.com

ABOUT AUTORICERCA

AutoRicerca is the journal of the *LAB – Laboratorio di Autoricerca di Base* (Laboratory of Basic Self-Research).

Its mission is to publish writings of value, mainly in Italian, on the topic of *all round research* (both inner and outer).

Standing outside the usual editorial categories, *AutoRicerca* offers to its readers articles of a high level, selected, translated and checked personally by the editor. These works, although they usually require some effort to be assimilated – they should be studied, more than read – remain nonetheless accessible to the willing general reader who is really eager to learn something new.

In accordance with the *Berlin Declaration*, which states that the dissemination of knowledge is only half complete if the information is not made widely and readily available to society, *AutoRicerca* is an *open access* journal.

More specifically, this means that the volumes in electronic format (pdf) are freely downloadable from the site of the *LAB*.

The open access to the electronic version does not preclude the possibility to order the paper volumes (one can also order a single volume), the purchase of which is a way to support the mission of the journal.

If you wish to be informed about the new releases (the actual cadence is of two issues a year), you can subscribe to the mailing list, by sending an email to the following address: *autoricerca@gmail.ch*, indicating in the object "mailing-list-journal," and specifying in the body of the message the name and country of residence.

PREVIOUS VOLUMES

NUMERO 1, ANNO 2011 – LO STATO VIBRAZIONALE

Un approccio alla ricerca sullo stato vibrazionale attraverso lo studio dell'attività cerebrale (*Wagner Alegretti*)

Attributi misurabili della tecnica dello stato vibrazionale (*Nanci Trivellato*)

Dal pranayama dello Yoga all'OLVE della Coscienziologia: proposta per una tecnica integrativa (*M. Sassoli de Bianchi*)

NUMERO 2, ANNO 2011 – FISICA E REALTÀ

Proprietà effimere e l'illusione delle particelle microscopiche (*Massimiliano Sassoli de Bianchi*)

Un tentativo di immaginare parti della realtà del micromondo (*Diederik Aerts*)

NUMERO 3, ANNO 2012 – L'ARTE DI OSSERVARE

L'arte dell'osservazione nella ricerca interiore (*M. Sassoli de Bianchi*)

NUMERO 4, ANNO 2012 – SCIENZA E SPIRITUALITÀ

Yoga, fisica e coscienza (*Ravi Ravindra*)

Cercare, ricercare, autoricercare...
Speculazioni su origine e struttura del reale (*M. Sassoli de Bianchi*)

NUMERO 5, ANNO 2013 – OBE

Scoprire la tua missione di vita (*Kevin de La Tour*)

Esperienze fuori del corpo: una prospettiva di ricerca (*N. Trivellato*)

Filtri parapercettivi, esperienze fuori del corpo e parafenomeni associati (*Nelson Abreu*)

Elementi teorico-pratici di esplorazione extracorporea (*Massimiliano Sassoli de Bianchi*)

NUMERO 6, ANNO 2013 – ENERGIA

Una sottile rete di luce (*Andrea Di Terlizzi*)

Bioenergia (*Sandie Gustus*)

Energie sottili o materie sottili? Una chiarificazione concettuale

Trasferimento interdimensionale di energia: un modello semplice di massa (*Massimiliano Sassoli de Bianchi*)

NUMERO 7, ANNO 2014 – SCIENZA, REALTÀ & COSCIENZA

Scienza, realtà e coscienza. Un dialogo socratico (*Massimiliano Sassoli de Bianchi*)

NUMERO 8, ANNO 2014 – ARCHETIPI

Astrologia elementale e aritmosofia (*Vittorio D. Mascherpa*)

La nuova astrologia (*Nadav Hadar Crivelli*)

Corrispondenze astrologiche: una prospettiva multiesistenziale (*Massimiliano Sassoli de Bianchi*)

NUMERO 9, ANNO 2015 – CORRISPONDENZE

Dialogando con Misha e Maksim (*autori anonimi*)

NUMERO 10, ANNO 2015 – STUDI SULLA COSCIENZA

Risultati preliminari sul rilevamento di bioenergia e dello stato vibrazionale mediante fMRI (*Wagner Alegretti*)

Requisiti per una teoria matematica della coscienza (*F. Faggin*)

Studi preliminari su evidenze di pseudoscienza in coscienziologia (*Flávio Amaral*)

Fisica quantistica e coscienza: come prenderle sul serio e quali sono le conseguenze? (*Massimiliano Sassoli de Bianchi*)

NUMERO 11, ANNO 2016 – CORRISPONDENZE BIS

Dialogando con Misha e Maksim… e alcuni altri (*autori anonimi*)

NUMERO 12, ANNO 2016 – DIALOGO SULLA REALTÀ

Tra mentore e pupillo. Dialogo sulla realtà / Between mentor an pupil. Talking about reality (*Massimiliano Sassoli de Bianchi*)
⇒ [ALSO AVAILABLE IN ENGLISH]

NUMERO 13, ANNO 2017 – DIALOGO SULLA MALATTIA

Tra mentore e pupillo. Dialogo sulla malattia (*M. Sassoli de Bianchi*)

NUMERO 14, ANNO 2017 – NDE

NDE – La prova della sopravvivenza (*Andrea Pasotti*)

NUMERO 15, ANNO 2018 – NDE

Lo Yoga Darshana di Patanjali
Elementi di Sadhana dello Yoga (*M. Sassoli de Bianchi*)

NUMERO 16, ANNO 2018 – DUE CUORI

Due cuori / Two hearts (*Massimiliano Sassoli de Bianchi*)
⇒ [ALSO AVAILABLE IN ENGLISH]

NUMERO 17, ANNO 2019 – SPUNTI DI OSSERVAZIONE

Spunti di Osservazione (*Antonella Spotti*)

NUMERO 18, ANNO 2019 – THE SECRET OF LIFE

The secret of life (*Diederik Aerts, Kigen William Ekeson Massimiliano Sassoli de Bianchi & Valéry Schneider*)
Quantum theory and conceptuality: matter, stories, semantics and space-time (*Diederik Aerts*)
Telos and Complexity (*Kigen William Ekeson*)
⇒ [ONLY AVAILABLE IN ENGLISH]

www.ingramcontent.com/pod-product-compliance
Lightning Source LLC
Chambersburg PA
CBHW030921180526
45163CB00002B/422